PHILOSOPHICAL REFLECTIONS ON DISABILITY

Philosophy and Medicine

VOLUME 104

For other titles published in this series, go to
www.springer.com/series/6414

PHILOSOPHICAL REFLECTIONS ON DISABILITY

Edited by

D. CHRISTOPHER RALSTON

Rice University, Houston, TX, USA

and

JUSTIN HO

*Berkeley School of Law, University of California
at Berkeley, Berkeley, CA, USA*

 Springer

Editors
D. Christopher Ralston
Rice University
Dept. Philosophy
6100 S. Main Street MS14
Houston TX 77005-1892
USA
ralston@rice.edu

Justin Ho
Berkeley School of Law
215 Boalt Hall
Berkeley, CA 94720-7200
USA
J.T.Ho@berkeley.edu

ISBN 978-90-481-2476-3   e-ISBN 978-90-481-2477-0
DOI 10.1007/978-90-481-2477-0
Springer Dordrecht Heidelberg London New York

Library of Congress Control Number: 2009929305

Cover design: Boekhorst Design b.v.

Printed on acid-free paper

Springer is part of Springer Science+Business Media (www.springer.com)

Preface

This project draws together the various strands of the debate regarding disability in a way never before combined in a single volume. The volume first of all seeks to offer a representative sampling of competing philosophical/theoretical approaches to the conceptualization of disability as such. This theoretical background serves as a crucial backdrop to the remainder of the book, which addresses such themes as (1) the complex interplay between disability and quality-of-life considerations, (2) questions of social justice as it relates to disability, and (3) the personal dimensions of the disability experience.

Consistent with these general themes, the primary goal of the volume is to bring together a collection of essays by important scholars in the fields of moral theory, bioethics, and disability studies to address such specific questions as the following:

- What is the best way to conceptualize disability or theorize about it? Should one adopt either the "medical model" or the "social model" of disability—or take a different approach altogether? What are the implications of adopting one model of disability versus another?
- Are there any identifiable connections between disability and reduced quality of life? Between disability and suffering? What are their moral implications? What should we think of such practices as prenatal testing for disability, and/or abortion on the basis of disability?
- What, if anything, is "owed" to persons with disabilities? Should they be "compensated" for their disability? "Repaired" so as to restore them to a "normal" level of functioning? Do the philosophically-dominant theories of social justice (e.g., egalitarian and consequentialist theories that tend to emphasize questions of distributive justice) offer sufficient resources for addressing the needs and problems faced by those with disabilities? Or should we look elsewhere (e.g., to libertarian theories, virtue-oriented theories, and the like) for the conceptual resources needed to adequately address questions of social justice and disability?

By explicitly locating the discussion of various *applied* ethical questions within the broader *theoretical* context of how disability is best conceptualized, the volume seeks to bridge the gap between abstract philosophical musings about the nature of disease, illness and disability found in much of the philosophy of medicine literature, on the one hand, and the comparatively concrete but less philosophical discourse frequently encountered in much of the disability studies literature. It also critically examines various claims advanced by disability advocates, as well as those of their critics. In this way, this volume is a unique contribution to the scholarly literature, and also offers a valuable resource to instructors and students interested in a text that critically examines and assesses various approaches to some of the most vexing problems in contemporary social and political philosophy.

Houston, TX, USA   D. Christopher Ralston
Berkeley, CA, USA   Justin Ho

Acknowledgements

We would like to extend our thanks to a number of individuals who were instrumental in bringing this project to fruition—in particular, the various contributors; H. Tristram Engelhardt, Jr.; and the entire Springer team. Special thanks also goes to Lisa Rasmussen for invaluable advice and guidance along the way.

Contents

Contributors

Ron Amundson Department of Philosophy, University of Hawaii, Hilo, Hawaii, USA, ronald@hawaii.edu

Christopher Boorse Department of Philosophy, University of Delaware, Newark, DE, USA, cboorse@udel.edu

Eric J. Cassell Department of Public Health, Weill Medical College, Cornell University, New York, USA, eric@ericcassell.com

Jean Bethke Elshtain Divinity School, University of Chicago, Chicago, Illinois, USA, jbelshta@uchicago.edu

H. Tristram Engelhardt, Jr. Department of Philosophy, Rice University, Houston, TX, USA, htengelh@rice.edu

John Harris Institute for Science, Ethics and Innovation, School of Law, University of Manchester, Manchester, UK, John.harris@manchester.ac.uk

Justin Ho Berkeley School of Law, University of California at Berkeley, Berkeley, CA, USA, J.T.Ho@berkeley.edu

Garret Merriam Department of Philosophy, University of Southern Indiana, Evansville, Indiana, USA, gamerriam@usi.edu

Lennart Nordenfelt Department of Medical and Health Sciences, Linköping University, Linköping, Sweden, lennart.nordenfelt@liu.se

Patricia M. Owens Department of Public Health, Weill Medical College, Cornell University, New York, USA, patowens@ericcassell.com

Laura M. Purdy Department of Philosophy, Wells College, Aurora, New York, USA, lpurdy@wells.edu

Muireann Quigley Institute for Science, Ethics and Innovation, School of Law, University of Manchester, Manchester, UK, Muireann.Quigley@manchester.ac.uk

D. Christopher Ralston Department of Philosophy, Rice University, Houston, TX, USA, ralston@rice.edu

Anita Silvers Department of Philosophy, San Francisco State University, San Francisco, California, USA, asilvers@sfsu.edu

Daniel P. Sulmasy, MacLean Center for Clinical Medical Ethics, University of Chicago, Chicago, Illinois, USA, dsulmasy@medicine.bsd.uchicago.edu.

Torbjörn Tännsjö Department of Philosophy, Stockholm University, Stockholm, Sweden, torbjorn.tannsjo@philosophy.su.se

Christopher Tollefsen Department of Philosophy, University of South Carolina, Columbia, South Carolina, USA, Christopher.Tollefsen@gmail.com

Robert M. Veatch Kennedy Institute of Ethics, Georgetown University, Washington, DC, USA, veatchr@georgetown.edu

Author Biographies

Ron Amundson, Ph.D. is Professor of Philosophy at the University of Hawaii at Hilo. His research is in the history and philosophy of evolutionary biology, and the concept of biological normality. His book *The Changing Role of the Embryo in Evolutionary Thought* was published by Cambridge University Press in 2005.

Christopher Boorse, Ph.D. is Associate Professor of Philosophy at the University of Delaware. He is best-known for his work on concepts of health and disease and on biological function. Besides philosophy of medicine and philosophy of biology, his research interests are in ethics and law.

Eric J. Cassell, M.D., M.A.C.P. is Emeritus Professor of Public Health at Weill Medical College of Cornell University and Adjunct Professor of Medicine at McGill University. His writings center on moral problems in medicine, the care of the dying and the nature of suffering. He is married to Patricia M. Owens, an expert on disability policy.

Jean Bethke Elshtain, Ph.D. is Laura Spelman Rockefeller Professor of Social and Political Ethics in the Divinity School at the University of Chicago. A political philosopher whose task has been to show the connections between our political and ethical convictions, she also holds appointments in the Department of Political Science and on the Committee on International Relations. She is the author, most recently, of *Sovereignty: God, State, and Self*, her Gifford lectures.

H. Tristram Engelhardt, Jr., M.D., Ph.D. is Professor of Philosophy at Rice University and Professor Emeritus at Baylor College of Medicine.

John Harris, D.Phil. is Lord David Alliance Professor of Bioethics at the School of Law of the University of Manchester. He is joint Editor-in-Chief of *The Journal of Medical Ethics*, the highest impact journal in medical and applied ethics. He is also Research Director for the Institute of Science, Ethics and Innovation, which focuses on the ethical questions raised by science and technology in the 21st century.

Justin Ho, M.A. is a law student at the Berkeley School of Law of the University of California at Berkeley.

Garret Merriam, Ph.D. is Assistant Professor of Philosophy at the University of Southern Indiana. He holds a Bachelor of Arts degree from the University of California-Davis, as well as a Master of Arts degree and a doctoral degree from Rice University. Prior to his appointment at the University of Southern Indiana, Merriam was employed as an assistant professor at Lone Star College in Kingwood, Texas.

Lennart Nordenfelt, Ph.D. is Professor of Philosophy of Medicine and Health Care in the Department of Medical and Health Sciences at Linköping University. He was also President of the European Society for the Philosophy of Medicine and Health Care (ESPMH) from 2001–2005. His principal scholarly contributions have been to action theory, theory of medicine and the theory of health and welfare; his theory of health draws upon concepts central to action theory.

Patricia M. Owens, B.S., M.P.A. is an expert on disability policy. She was formerly the Associate Commissioner of Social Security, responsible for the United State's Social Security Disability Programs. She has been a senior officer in the private disability insurance sector, is an advisor on disability issues to both governmental and private organizations, and serves on the Board of the National Academy of Social Insurance.

Laura M. Purdy, Ph.D. is Professor of Philosophy and Ruth and Albert Koch Professor of Humanities at Wells College. Her research encompasses bioethics, applied ethics, and political philosophy, with particular emphasis on feminist bioethics. Her work looks especially at new technologies and issues concerning families and children, focusing on the ethics of sexuality and reproduction.

Muireann Quigley, B.Sc., MB Ch.B., M.A. is a medical doctor and Lecturer in Bioethics in the Centre for Social Ethics and Policy, and the new Institute for Science, Ethics, and Innovation, at the University of Manchester. She previously worked as a Research Fellow in Bioethics and Law at the University of Manchester. She is working toward a Ph.D. degree, with a focus on the question of property rights in human tissue.

D. Christopher Ralston, M.A., Ph.D. (cand.) is a doctoral candidate in the Department of Philosophy at Rice University and an Assistant Managing Editor of the *Journal of Medicine and Philosophy*. His research interests include ethics, bioethics, and the philosophy of medicine. He is currently writing a dissertation on concepts of disability.

Anita Silvers, Ph.D. is Professor of Philosophy at San Francisco State University.

Daniel P. Sulmasy, O.F.M., M.D., Ph.D. is the inaugural Clinton-Kilbride Professor of Medicine and Medical Ethics, Professor of Divinty, and Associate Director of the MacLean Center for Clinical Medical Ethics at the University of Chicago. He previously held the Sisters of Charity Chair in Ethics at St. Vincent's Hospital–Manhattan and served as Professor of Medicine and Director of the Bioethics Institute of New York Medicine College. He was appointed by

Gov. Pataki to the New York State Task Force on Life and the Law in 2005. He serves as Editor-in-Chief of the journal *Theoretical Medicine and Bioethics*, and is the author of four books: *The Healer's Calling* (1997), *Methods in Medical Ethics* (2001). *The Rebirth of the Clinic* (2006) and *A Balm for Gilead: Meditations on Spritiuality and the Healing Arts* (2006).

Torbjörn Tännsjö, Ph.D. is Kristian Claëson Professor of Practical Philosophy at Stockholm University, director of Stockholm Bioethics Centre, and Affiliated Professor of Medical Ethics at the Karolinska Institutet. During the fall of 2008 he was a research fellow at The Swedish Collegium for Advanced Study. He is a Member of the medical ethics committee of The National Board of Health and Welfare (the Swedish Government agency responsible for the supervision, evaluation and monitoring of social services, health care and medical services, dental care, environmental health, and control of communicable diseases).

Christopher Tollefsen, Ph.D. is Professor of Philosophy in the Department of Philosophy at the University of South Carolina. He has published articles in ethical theory and medical ethics and is the series editor of Springer's Catholic Studies in Bioethics.

Robert M. Veatch, Ph.D., is Professor of Medical Ethics and the former Director of the Kennedy Institute of Ethics at Georgetown University, where he also holds appointments as Professor of Philosophy and Adjunct Professor in the Department of Community and Family Medicine at Georgetown Medical Center. He is the Senior Editor of the *Kennedy Institute of Ethics Journal* and a former member of the Editorial Board of the *Journal of the American Medical Association*. He served as an ethics consultant in the preparation of the legal case of Karen Ann Quinlan, the woman whose parents won the right to forego life-support (1975–76) and testified in the case of Baby K, the anencephalic infant whose mother insisted on the right of access to ventilatory support.

Chapter 1
Introduction: Philosophical Reflections on Disability

D. Christopher Ralston and Justin Ho

The discussions in this volume take place against the backdrop of the development of an increasingly vocal "disability rights" movement (henceforth DR) and the voluminous scholarly output of that movement's "theoretical arm" (Snyder, 2006, p. 478), the growing academic field of "disability studies." Since some readers may be unfamiliar with these movements and their relationship to what might be termed "standard" or "mainstream" bioethics, a brief historical overview will help to set the stage for understanding the various disputes to which the essays in this volume attend.[1] For purposes of this sketch, it will be helpful to think in terms of three major "eras" of the modern DR movement.

The beginnings of the first "era" of the contemporary DR movement can be traced to the early 1960s. Prior to that time, disability had generally been considered a "problem" falling exclusively (or at least primarily) under the purview of medical and/or rehabilitation professionals. Individuals with disabilities were typically referred to these professionals either for "cure" or for "rehabilitation," where the focus was on enabling individuals with disability to adjust to the society in which they lived. With the advent of large-scale entitlement programs such as Medicare and Medicaid, these professionals also increasingly took on a "gate-keeping" role (Snyder, 2006, p. 481), determining whether individuals qualified for benefits under those programs. Research on disability, and the development of public policy with respect to disability, rarely relied upon the perspectives or expertise of disabled persons themselves. Overall, social responses to disability emphasized charity toward and/or protection of those with disabilities—if not outright elimination, as suggested by well-respected proponents of the eugenics movement that reached its heyday in the early twentieth century (Snyder, 2006, p. 481; see also Snyder & Mitchell, 2006).

In response to these historical trends, disability advocates increasingly insisted during the 1960s and 1970s (and beyond) that reflection and research on disability, the development of social policies regarding disability, and so forth, ought to be conducted by disabled persons themselves (cf. Charlton, 2000). To facilitate this

D.C. Ralston (✉)
Department of Philosophy, Rice University, Houston, TX, USA
e-mail: ralston@rice.edu

D.C. Ralston, J. Ho (eds.), *Philosophical Reflections on Disability*, Philosophy and Medicine 104, DOI 10.1007/978-90-481-2477-0_1,
© Springer Science+Business Media B.V. 2010

objective, the DR movement advocated for the establishment of a new academic field, one that would eventually come to be called "disability studies." The DR movement also advocated a shift from "demands for charity" on behalf of persons with disability, to a focus on "demands for civil rights" (Burgdorf, 2006, p. 94). In this respect, the DR movement saw itself in continuity with the broader civil rights movement of the era, drawing inspiration from the accomplishments of that movement, including the passage of the Civil Rights Act of 1964.

Early precursors of what would eventually become a full-blown DR movement included, in the 1960s, the "person first movement," which "sought to upgrade social awareness by interrogating the linguistic implications of referring to persons with disabilities as handicapped, crippled, or disabled [rather than as *persons with disabilities*]" (Snyder, 2006, p. 483); and, in the 1970s, the "independent living movement," which advocated for the deinstitutionalization of persons with disabilities who previously would have been confined to nursing homes and other similar institutions, in many cases for their entire lives. Notably in this context, the Union of the Physically Impaired Against Segregation (UPIAS) formulated, in the first half of the 1970s, what was likely the first explicit articulation of what would later come to be termed the "social model" of disability—defining disability as "the relationship between people with impairments and a society that excludes them" (Shakespeare, Bickenbach, Pfeiffer, & Watson, 2006, p. 1103).

The 1980s constitute a second major "era" in the development of the DR movement. During this period the DR movement urged a rejection of so-called "medical model" or "cure" approaches to disability, favoring instead an approach that emphasized a reversal of the required direction of "adjustment"—that is, instead of requiring that individuals with disability adjust to their surrounding society, DR activists insisted that *society* ought to adjust itself to persons with disabilities (e.g., by removing environmental barriers and enhancing the social inclusion of those with disabilities [Shakespeare et al., 2006, p. 1105]). In this context, a "first generation" (Snyder, 2006, p. 481) of disability activist-scholars developed a broad array of new theoretical approaches to disability. In addition to the "strict social model" (Snyder, 2006, p. 485) of the sort that had been articulated by UPIAS in the 1970s, this first generation of activist-scholars also developed other ways of conceptualizing disability, including the "minority group model of disability," according to which disabled persons constitute a distinct, and oppressed, minority; the notion of "disability as culture," according to which particular groups of disabled persons (e.g., those who are deaf) constitute a discrete culture and possess a unique cultural heritage; and the idea that disability can be conceptualized as a "diversity category" similar to race, gender, or sexual orientation (Snyder, 2006, p. 485). In this way, the nascent field of "disability studies" came to bear affinities with other new disciplines that also drew upon notions of identity politics, such as queer studies, gender studies, and the like (Snyder, 2006, p. 485).

The most recent "era" of the DR movement stretches roughly from 1990 to the present. Since the passage in 1990 of the Americans with Disabilities Act (ADA), a second "generation" (Snyder, 2006, p. 484) of activist-scholars has continued to develop various theoretical approaches to disability, and has also sought to engage with "standard" or "mainstream" bioethics on questions of interest to the DR movement. Over time, this dialectical relationship—one in which DR draws upon and interacts with, often in criticism, the resources of mainstream academic bioethics—has resulted in some noticeable divergences between the movements, despite similarities in concern and focus.[2] Most significantly, fundamental differences between the DR movement and "bioethics" tend to manifest themselves in at least two areas: their "understanding" of the nature of disability, and resulting differences in their respective "valuations" of lives with disabilities (Asch & Wasserman, 2006, pp. 165–166). As Asch and Wasserman explain, DR advocates believe there to be a "gap in understanding" between the two movements, a gap rooted in a "fundamental misunderstanding of the nature of disability" (Asch & Wasserman, 2006, p. 166). This misunderstanding is, in turn, rooted in "standard" bioethics' embrace of a "medical model assumption"—namely, that "functional impairment is the sole or primary cause of what is presumed to be an unacceptable, unsatisfying life" (Asch & Wasserman, 2006, p. 166). The facile embrace of this assumption is, DR advocates contend, the result of a prior acceptance of two further, "erroneous" assumptions— first, the assumption that life with a chronic disability is *permanently* disrupted in the same kind of way that life is disrupted by a flu or other systemic illness; and second, an assumption that the isolation, poverty, and other negative sequelae often associated with disability are the "inevitable consequences of biological limitation" (Asch & Wasserman, 2006, p. 166). Needless to say, DR advocates strongly dispute both of these assumptions and counter, instead, that life with a disability is not an "unremitting tragedy," and that the *real* problem in living life with a disability is the surrounding social, institutional, and physical environment with which persons with disabilities must deal (Asch & Wasserman, 2006, p. 166).

Given these prior assumptions, DR advocates argue, the two movements tend to arrive at fundamentally different assessments of the quality of life with disabilities. These differences between the movements manifest themselves in a number of concrete bioethical contexts, including (1) the creation and extension of lives, e.g., with respect to questions about prenatal testing and selective abortion for disability, as well as questions surrounding the medical treatment of impaired newborns; (2) life and death decision making (e.g., in differing understandings of the meaning and significance of such notions as "dependence," "independence," and "interdependence," with implications for controversial practices such as physician-assisted suicide); and (3) justice in healthcare allocations (with questions, e.g., about whether or not—and to what extent—the presence of disability should affect the types and/or extent of medical treatment received) (Asch & Wasserman, 2006, pp. 167–170).

As the foregoing suggests, discussions about disability tend to cluster around three major themes: (1) how is disability best *defined* or *conceptualized*?; (2) what impact does disability (however defined) have on persons with disabilities, particularly in terms of their "quality of life"?; and (3) how ought we to *respond* to disability, either at the individual level (questions of bioethics) or at the social level (questions of public policy and social justice)? The essays in this volume touch on these and other related issues; accordingly, the volume is organized loosely around these three thematic categories.

This volume first seeks to offer a representative sampling of competing philosophical/theoretical approaches to the conceptualization of disability as such. By explicitly locating the discussion of various *applied* ethical questions within the broader *theoretical* context of how disability is best conceptualized, the volume seeks to bridge the gap between abstract philosophical musings about the nature of disease, illness and disability found in much of the philosophy of medicine literature, on the one hand, and the comparatively concrete but less philosophical discourse frequently encountered in much of the disability studies literature. It also critically examines various claims advanced by disability advocates, as well as those of their critics. In this way, the volume is a unique contribution to the scholarly literature, and offers a valuable resource to instructors and students interested in a text that critically examines and assesses various approaches to some of the most vexing problems in contemporary social and political philosophy.

1.1 The Concept of Disability

An adequate philosophical discussion concerning disability can be carried out properly only if one has a firm understanding of what disability is. However, as the authors in this section illustrate, there are numerous ways of defining disability.

As Anita Silvers notes in "An Essay on Modeling: The Social Model of Disability," there have been traditionally two different types of models that have been given to show (1) what disabilities are and (2) the origin (or cause) of disabilities. Broadly speaking, those who adhere to the *medical model* of disability hold that the harm associated with disability is the result of something (a defect, flaw, etc.) inherent in the person who has the disability; those who adhere to the *social model* of disability understand this harm to be the product of unjust social structures.

Silvers acknowledges that some argue that the social model fails to adequately account for the suffering that many persons with disabilities encounter in their daily lives or advances an ideal of independence that "some disabled people's dysfunctions make unrealizable for them," and that the medical model can better account for these phenomena. Nevertheless, Silvers states that the virtues of the social model are substantial. For example, merely claiming that disabilities result from defects in the physical structures of the individual fails to capture the fact that fact many of those with disabilities are able to participate successfully in society when certain measures have been taken to alter the physical or social environment. She also notes that persons with the same medical condition may not be considered disabled; whether or not they are considered "disabled" depends in part on the goals they pursue. Furthermore, whichever definition one adheres to will have practical consequences in regard to social policy. If one claims that disability is the product of something inherent in the individual, then efforts should be made to correct the individual; on the other hand, if the harm associated with disability is due solely to society, then persons with disabilities are best seen as members of a minority whose rights are being violated by an unjust majority. If this latter notion is taken seriously, then the focus of remedial action will be on enabling persons with disabilities to achieve greater freedom of participation in social life and opportunities. Silvers concludes her essay by claiming that different models of disability are appropriate for different values and goals, and suggests that both may have their place in a pluralistic society.

It is clear from reading Silver's essay that she believes that concepts of disability necessarily have some evaluative component. By adopting certain concepts, one is in a sense endorsing certain moral values, goals, or principles. One question readers might ask is whether (1) it is even possible to create a value-free concept of disability and, if so, (2) what rationale might be given for endorsing such a concept. Furthermore, if one holds (as Silvers does) that there is a moral rationale for adopting certain models of disabilities in certain contexts, then one needs to raise the question whether these moral goals would best be achieved by utilizing the existing social and medical models of disability. For example, one might wonder whether a model that combines elements of both of these models might be more advantageous than keeping the two models distinct and simply applying them in different situations.

Lennart Nordenfelt briefly sketches out what such a hybrid model of disability might look like in "Ability, Competence and Qualification: Fundamental Concepts in the Philosophy of Disability." Nordenfelt spends much of his paper examining the concept of positive ability in order to show, by contrast, what types of *non-abilities* exist before finally giving an account of which non-abilities are *disabilities*. He approaches this topic by focusing specifically on the ability (or non-ability) to work professionally. According to Nordenfelt, to have the ability to work, one must have the following conditions:

1. Overall competence for the job, including knowledge and skill.
2. Toleration of physical, psychological, and social aspects of the job.
3. Courage with regard to taking up the job and fulfilling the demanding tasks of the job.
4. Virtues necessary for fulfilling the tasks of the job.
5. Qualifications for having and performing the job.
6. Executive ability to perform the job.
7. Willingness to take and perform the job.

Bearing this in mind, he claims that not all inabilities are disabilities. Rather, a disability is a non-ability to realize one or more of one's vital goals given standard or accepted circumstances. Furthermore, he states that disabilities are the product of one or more factors that are inherent in the agent *in conjunction* with the environment.

The concept of disability that Nordenfelt provides suggests that both the medical and the social model of disability, while not inaccurate, are incomplete; his approach thus represents an attempt to meld together the causal dimensions of both these models. However, while his account clearly has normative implications, Nordenfelt does not mention these in his article. The reader might want to consider whether his account not only has the virtue of being more accurate, but whether it also is more advantageous in terms of its practical implications. One of the reasons why there is opposition to both the social and the medical model is that for many it is counterintuitive to *always* alter the environment or *always* alter one's body in order to ameliorate or eliminate disabilities. If disability is presented as the product of *both* one's environment *and* features of the person, one might then argue that it may be appropriate in some cases to "fix" one of these causal conditions as opposed to the other, depending (in part) on pragmatic considerations. This in turn might alleviate some of the controversy that stems from trying to advocate only the medical or the social model.

In this section's final essay, "Disability and Medical Theory," Christopher Boorse analyzes the concept of disability by comparing it with the fundamental concepts of medicine, particularly those of medical theory. He approaches this subject from three different angles: (1) conceptual analysis, with particular attention given both to common-sense usage of the term and its usage by specialists; (2) legal analysis, with special attention paid to its use in the Americans with Disabilities Act (ADA); and (3) disability advocacy, with a comparison of two distinct types of approach to such advocacy. Cumulatively, the discussion of these three domains yields four central conclusions, which can be summarized, in their barest terms, as follows:

(1) Conclusion #1: There is no single, univocal disability concept—i.e., it is a "highly indeterminate concept," such that, "in practice, there is no one disability concept."
(2) Conclusion #2: Disability is a species of "gross impairment," but gross impairment is not identical to disability—i.e., "gross impairment" is a broader concept than disability.
(3) Conclusion #3: There is no simple relationship between "disability" and "disorder"/"pathology." In particular,

    a. Conclusion #3a: "…'impairment' means nearly the same as 'clinically evident pathological condition.'"
    b. Conclusion #3b: "…an impairment is neither sufficient nor necessary for a disability."

(4) Conclusion #4: "disability" is not a purely medical concept—i.e., the concept of disability includes both medical and non-medical components.

In support of conclusion #1, Boorse argues that the term "disability" has no "clear ordinary meaning," demonstrating through textual analysis that there is variation both in common-sense usage of the term and in its usage among professionals in the field. In support of conclusion #2, Boorse argues that the claim that disability is equivalent to gross impairment is problematic because (a) it allows too much to be counted as "disability," (b) it would allow *any* species to be considered "disabled," and (c) it fails to allow conceptual room for "normal disability," as a result of which we may want to reject the equation. In arguing for conclusion #3, Boorse analyzes the ADA and the WHO statements on disability, showing that both uses of the term (i.e., conclusions #3a and #3b above) are evident in these documents. Finally, in support of conclusion #4, Boorse shows that "disability" must be analyzed in terms of at least three or four variables, which in turn determine the meaning of the term. As Boorse puts it, "'*x is disabled by impairment I*' really means something like '*Because x has impairment I, which significantly limits x in activities of type A in environments of type E, x deserves the consequence C.*'" Boorse uses this contextual analysis of disability to show that: (a) of two medically identical people, one may be disabled, another not; and (b) a "paradigm" of disability in context A may *not* be disability in context B.

By way of analysis, it is important to note, first of all, that Boorse's argument is largely devoted to a descriptive account, arrived at by means of detailed textual analysis, of the various ways in which the term "disability" is *used*, rather than to developing a full-blown *theory* of disability. Boorse's ultimate aim, however, is to connect current uses of the term with the fundamental terms and concepts of medical theory, with a view toward demonstrating the relevant similarities and dissimilarities among them. Second, Boorse's argument in this paper is not motivated by any particular theory of justice; in fact, his argument is arguably compatible with a number of different theories of justice. One might, for example, use Boorse's analysis as a foundation upon which to argue in a variety of directions when it comes to specific ethical questions (e.g., questions related to whether medical "treatment" for disability is warranted), as Boorse himself acknowledges in the concluding section of his paper. Third, it is important in this context to distinguish between a *causal* model of disability (what *causes* disability?) and what we might call an *identification* model of disability (how do we *identify* disability?). Whereas Silvers and Nordenfelt appear to combine these two types of models in their account of disability, Boorse appears to be giving us only an identification model; indeed, in endnote 12, he explicitly eschews "all metaphysical issues about disability, including causation, and all empirical causal questions as well." Here again, the reader will want to consider whether it is possible—and if so, whether it is desirable—to separate causal and identification-related claims about disability in this way.

1.2 Disability, Quality of Life, and Bioethics

Much of the philosophical literature on disability has focused on the issue of the quality of life of those with disabilities. For example, the assumption that having a disability in some way negatively affects the quality of a person's life plays a crucial role in many of the arguments which claim that prenatal screening, selective abortion, and gene therapy are permissible, as such interventions may prevent the birth of persons with disabilities or may be used to eliminate certain disabilities altogether. Such arguments, of course, all assume that the quality of life of persons with disabilities can in some way be judged or measured.

In "Utilitarianism, Disability, and Society," Torbjörn Tännsjö employs classical hedonistic utilitarianism to determine what social policies should be adopted in regard to persons with disabilities. Like all classical hedonistic utilitarians, Tännsjö is interested in what social arrangements will maximize the sum-total of happiness in the world. For utilitarians like Tännsjö, the capacity to experience happiness is the single most important consideration in determining what our moral obligations are to persons with disabilities. Among some of the interesting claims that Tännsjö makes are that empirical research suggests many persons with disabilities can live very happy lives.

This utilitarian framework also leads Tännsjö to make a number of controversial claims. For example, in response to the question as to whether society should allow the use of prenatal genetic diagnosis and selective abortion, Tännsjö, like many utilitarians, claims that such procedures should be allowed, as they prevent persons from existing who would otherwise diminish the sum total of happiness. However, such procedures should not be obligatory, according to Tännsjö, since such a policy might threaten the well-being of people currently living with disabilities.

For many, Tännsjö's arguments will not be convincing, as his arguments assume that happiness is quantifiable and that we have access to a metric that allows us to accurately determine which actions best promote happiness. Others may argue that this framework produces ethical judgments and normative prescriptions that are abhorrent on their face, and for this reason alone such a theory should be dismissed. Still others might simply argue that hedonistic utilitarianism is not inherently flawed, but must be balanced against other ethical appeals such as the appeal to justice and respect for persons.

In "Too Late to Matter? Preventing the Birth of Infants at Risk for Adult-Onset Disease or Disability," Laura M. Purdy argues that it is sometimes permissible to prevent the birth of persons with disabilities either by avoiding conception or by prenatal screening followed by abortion. Purdy uses the examples of breast cancer and Huntington's Disease to illustrate (1) that some diseases which might also be thought of as disabilities may severely undermine the quality of life of the persons who possess them and (2) that these are cases in which it may be morally appropriate to prevent the birth of individuals who have or will likely develop these diseases.

Purdy goes on to dismiss what she believes are the assumptions which underlie the arguments of those who reject such measures. She attends first to the premise that some possible—or, in her terms, "phantom"—child has a right to be brought into existence. Purdy claims that it is metaphysically absurd to attribute rights to possible but not actual entities. She also calls into the question the view held by many advocates of the social model of disability that there is nothing intrinsically bad about disabilities, arguing that, were we to discover a vaccine for Huntington's Disease and other disabling diseases, we would have no problem with utilizing such interventions. This, according to Purdy, (1) suggests that there is something inherently wrong with disabilities, and (2) calls into question the claim that other means of preventing the birth of persons with disabilities should be regarded as problematic.

It is interesting to note that Purdy does not seem to find there to be a morally relevant difference between not conceiving children with disabilities, on the one hand, and selective abortion on the other. This is in part because she does not regard fetuses as having the same morally relevant status as adult humans, as a consequence of which they do not have a right not to be killed. This is of course a very controversial claim that has divided the bioethics community. Suppose Purdy restricted her discussion of preventing births of children with disabilities to the claim that some persons should not conceive children at all, and then used the same form of argumentation. Would her arguments still seem so objectionable? It is worth considering this question, as some might argue that her assumptions regarding selective abortion may in some way prejudice how one assesses her arguments about prenatal screening and avoiding conception to prevent the birth of children with disabilities. Finally, one should consider whether changes in social attitudes and structures might significantly improve the quality of life of many of those with disabilities, and to what extent (if any) such external remedies might also be applied beneficially to those with breast cancer and Huntington's Disease.

In "To Fail to Enhance is to Disable," Muireann Quigley and John Harris argue for a definition of disability that does not appeal to any notion of normalcy. They claim that a disability is any physical or mental condition that, *ceteris paribus*, a person would rationally prefer not to have because it harms them. A harming condition is in turn defined as any condition which thwarts a person's interests, and failing to implement a remedy when one is readily available is a necessary condition of the harm. With this definition in place Quigley and Harris want to argue that parents are not only causally but also morally responsible for any disability that a child develops, if it is possible to enhance children so they do not develop such disabilities and parents are aware of the existence of such interventions.

Harris and Quigley also argue that any distinction between treatment and enhancement is arbitrary by pointing to such examples as immunizations. Therefore, they claim, opponents of enhancement cannot appeal to the treatment-enhancement distinction in order to justify the impermissibility of enhancement. Moreover, Harris and Quigley argue, the opponents of enhancement assume that categorizations are morally relevant in themselves, when what should be relevant is the relative harm or benefit that results from the procedure in question.

The definition of disability offered by Quigley and Harris is in many ways similar to that offered by Nordenfelt. Disability has a subjective dimension: whether impairment is a disability is relative to the goals of the bearer of the disability. It is important to note that the authors assume that it is possible to harm someone through an omission. Such claims are not unusual in the philosophical literature. Nevertheless, the reader may question whether it is appropriate to think that parents are harming their children if they do not intend to thwart their children's interests, despite being aware of the availability of these procedures. One reason to think that parents might still harm their children in such circumstances has to do with the notion that along with certain roles come specific responsibilities and duties. One such duty may be the duty to act in the best interests of one's child or the duty to try to give one's child the best life that they can.

Finally, in "Rehabilitating Aristotle: A Virtue Ethics Approach to Disability and Human Flourishing," Garret Merriam develops an updated Aristotelian moral theory to illumine our understanding of disability and its relationship to human flourishing. However, unlike Aristotle, Merriam does not couch human flourishing in any sort of species essentialism, holding that such a view is both arbitrary and displays a form of chauvinism. For Merriam, whether a particular individual flourishes or not is contingent on whether she is living well given the individual circumstances of her life. Merriam's framework leads to a number of interesting conclusions. Persons like Helen Keller are described as the paradigm of human flourishing, whereas persons with anencephaly fall on the opposite end of the spectrum. Merriam's theory also yields conclusions that many might find counterintuitive; for example, he argues that in some contexts it is morally acceptable for deaf couples to take steps to ensure that their children are deaf.

Merriam's paper raises a number of important questions. Many people rely on general claims about the quality of life of persons with disabilities or moral status to argue for or against the use of certain medical procedures or policies. However, many of these general claims are either based on atypical examples or, conversely, fail to take into account marginal cases. By contrast, Merriam's approach forces us too look very closely at the individuals who will be affected by our actions and tailor our moral judgments to the specific circumstances surrounding their lives. It is important for the reader to assess whether such an approach escapes the problems associated with many of the other arguments which invoke some notion of quality or moral status in determining what polices should be adopted in regard to those with disabilities.

1.3 Disability and Justice

Besides issues involving the quality of life, the other most frequently discussed topic in the philosophical literature on disability is the issue of what obligations society has towards individuals with disabilities. For example, does the fact that one has a disability provide a morally compelling reason for society to remedy his condition? And if the answer to this question is "yes," what sort of efforts should be employed to achieve this goal? Should such efforts take the form of compensation or integration? Or does it also involve cultivating changes in the attitudes of those persons without a disability?

Furthermore, even if we accept that society is obligated to aid those with disabilities, what do we do when those duties conflict with the important interests of the non-disabled majority? Do the interests of the non-disabled sometimes serve as a constraint or trump the obligations that society has towards those with disabilities? Each of the authors in this section attempts to answer all or some of these questions. However, as shall become obvious, each approaches these questions from a different moral perspective.

In "Equal Treatment for Disabled Persons: The Case of Organ Transplantation," Robert M. Veatch assumes the truth of an egalitarian theory of justice, which requires that we treat equally persons who possess the same morally relevant properties. Veatch also assumes the truth of the "difference principle," which claims that a society should concentrate its resources on benefiting those who are among the least well off. However, an important question for persons who hold both these views is whether having a disability provides a basis for different treatment, as chronic disabilities may sometimes require an enormous expenditure of resources with little benefit. Veatch attempts to answer this question using the example of organ transplantation.

Veatch ultimately concludes that publicly held organs may justly be withheld from those with permanent total unconsciousness. His reasoning is that persons who are permanently unconscious are not as bad off as those who are conscious and so the interests of those who are conscious have priority. Nevertheless, such organs should only be withheld from those with less-severe disabilities on the same grounds that apply to all candidates for organ transplant. However, the presence of certain disabilities may indirectly impact allocation priority, since the disabilities in question may affect whether or not individuals have *other* morally relevant characteristics. So there is still a sense in which disabilities appear to impact the allocation of resources.

When evaluating Veatch's argument, it is important to assess whether he ultimately succeeds in defending both of these principles. Some readers, for example, might think that there is tension between the conclusions that Veatch draws concerning "lesser disabilities" and the "difference principle." As Veatch acknowledges, persons who are considered good candidates for organ transplantation are usually not among the worse off, and *prima facie* the difference principle seems to favor persons who are poor candidates. Hence, there may be a mismatch between the prescriptions endorsed by the difference principle, on the one hand, and the egalitarianism that Veatch promotes, on the other. Some might hold that in trying to specify the difference principle to account for such cases as Veatch does, the essence of the difference principle is ultimately lost and replaced by egalitarian considerations.

In "Disability Rights: Do We Really Mean It?" Ron Amundson argues for the claim that the attitudes expressed towards those with disabilities are often inappropriate, and that they can be seen in the continued failure of many mainstream academics to grant to the disability rights movement the legitimacy that is attributed to other civil rights movements. In making this argument, Amundson focuses on *From Chance to Choice* (Buchanan, Brock, Daniels, & Wikler, 2000), a highly acclaimed scholarly work on bioethics, in an effort to show that many academics openly express demeaning attitudes towards those with disabilities—attitudes which would not be acceptable were they to be expressed toward other disadvantaged groups. For example, the authors of *From Chance to Choice* claim that integrating persons with disabilities into society is in many cases "unduly burdensome to others" and that the dominant group (which in this case is the non-disabled) has a morally legitimate interest in maintaining segregationist policies. Amundson argues that such comments would be decried if they were made in regard to racial minorities and women. However, in the case of those with disabilities, such comments are accepted without impunity.

In making his argument, Amundson demonstrates that justice involves more than the distribution of goods and services. Treating persons justly also entails showing them the respect that they deserve, and one way to show such respect is by displaying the appropriate attitudes towards persons. If we assume as Veatch and many others do that persons who share the same morally relevant properties should be treated the same, then it is necessary for the non-disabled to work to alter their discriminatory attitudes towards persons with disabilities and acknowledge that they are no different than any other disadvantaged group.

In "Dignity, Disability, Difference and Rights," Daniel P. Sulmasy quickly dismisses what he calls "The Standard Civil Rights Formula," which has been employed by various civil rights movements to argue for the right to equality of opportunity in society. He observes that this approach assumes that there are no morally relevant differences among human beings, an assumption that is difficult to justify. He then goes on to try to show that the basis for disability rights is human dignity.

Sulmasy argues that it is in virtue of being a human being that all humans have intrinsic value or dignity. To be human is to be a member of natural kind, which has various capacities to perform certain activities, which activities we regard to be intrinsically valuable. Disabilities, in turn, occur when diseases and injuries diminish those capacities that make an entity a member of a natural kind and play a role in the flourishing of that kind. It follows, then, that persons with disabilities have the same moral status as other human beings and are thus intrinsically valuable. Furthermore, because persons with disabilities are intrinsically valuable, and because their disabilities prevent them from flourishing, they have a right to those things which will promote their flourishing. However, Sulmasy cautions that positive rights are correlated (only) with imperfect duties. Therefore, there may sometimes be limits to what goods should be provided to those with disabilities. For example, persons with disabilities do not have an unlimited right to as much healthcare as they need, if providing such healthcare would jeopardize the overall well-being of society as a whole.

It is important to note that many philosophers might reject Sulmasy's account of moral status, given that he ultimately grounds moral status in the property of being human. For example, many of those who hold that animals have the same or greater moral status as some humans might hold that Sulmasy is wrong to think that properties that are inherent in a type also extend to each of its members. Such persons might claim that an individual's moral status depends on whether it possesses certain qualities and not on whether it belongs to a group which possesses those qualities.

In "Public Policy and Personal Aspects of Disability," Patricia M. Owens and Eric J. Cassell examine how the categorical definitions of disability that underlie the disability programs in the United States are morally inadequate because they fail to consider the social dimension of disability. Like some of the other authors in this volume, Owens and Cassell note that one's medical condition is not enough to disable someone. Like Silvers, they point out that different persons with the same medical condition may not be considered disabled, and that it is also possible for certain disabilities to disappear when the social context changes. To remedy these deficiencies in current disability programs, Owens and Cassell argue that society needs to take into account what it means to be a "person" when developing its policies towards persons with disabilities. To be a person is not only to be a free being but to be a being with a unique history, who stands in and is capable of having valuable relationships with other beings. Bearing this in mind, Owens and Cassell claim that ethical public policies can best be developed by taking "an individualized approach to evaluating function and then assigning responsibility for improving and maximizing function through appropriate personal and societal interventions."

By way of comparison, it is worth noting that, like Merriam, the authors argue that an individualized approach to public policy is the most ethical, and they offer a brief sketch of what such an effort would require. However, Merriam's approach differs from Cassell and Owens' in several important respects. Whereas Merriam couches his approach in notions of human flourishing, Cassell and Owens appear to favor a more contractarian approach. They argue that all the relevant parties should come together and engage in a discussion in which all the pertinent goods are identified and a policy is jointly adopted which best takes into account these goods.

In "Disability and Social Justice," Christopher Tollefsen offers a theory of justice which, he believes, overcomes the inadequacies associated with liberalism. Like many philosophers, Tollefsen argues that liberal theories of justice such as the one advanced by John Rawls fail to adequately take into account the interests of those who are disabled (and their caretakers), because they are based on some notion of a social contract. Such theories must assume that the persons who enter into such a contract are independent rational beings, and of course not all persons with disabilities possess these qualities. In contrast, Tollefsen argues for a theory of justice that is grounded in natural law. He begins with the claim that human flourishing requires the presence of certain goods, many of which can only be realized through social cooperation and an overarching political authority. Bearing this in mind, both persons with disabilities and their caretakers have needs for human flourishing that they cannot meet without the presence of certain social structures. If one assumes, as Tollefsen does, that persons with disabilities have the same moral status as those without disabilities, then efforts should be made to promote the flourishing of these individuals and their caretakers while also taking into account the needs of the public at large.

Tollefsen's article connects with many themes found in the other articles in this volume. Like Sulmasy, he argues that persons with disabilities have the same moral status as those without such impairments, and like Amundson he assumes that justice also involves displaying certain attitudes towards such persons. Finally, like Merriam, he assumes that persons with disabilities can also partake in human flourishing. However, there are important differences. Unlike Sulmasy he assumes that persons with disabilities have *all* the capacities that are typical of human beings, and therefore have the same moral status; disabilities merely prevent such capacities from being realized. Clearly, the soundness of Tollefsen's account of moral status depends largely on what it means to have a capacity. And unlike Merriam, he seems to hold that *all* persons with disabilities are capable of human flourishing. Finally, Tollefsen states that the interests of persons with disabilities must be reasonably balanced against the interests of those without disabilities. However, he does not offer any theory to help us resolve conflicts of interests when they arise. This is, of course, a problem with which most theories of justice struggle. If one appeals to merely utilitarian considerations, then the interests of persons with disabilities may often be overshadowed by the non-disabled majority. Without going into too much detail here, morally sound judgments will require comparing the *content* of interests as opposed to merely trying to compare the *numbers* of interests. For example, the interest in having an equal opportunity to fully utilize one's talents may outweigh the interest of not wanting to pay a few more cents in taxes.

Finally, in "The Unfair and the Unfortunate: Some Brief Critical Reflections on Secular Moral Claim Rights for the Disabled," H. Tristram Engelhardt, Jr., challenges the claim that secular morality can provide a justification for disability rights. It is commonly held that considerations of justice provide a rationale for why society is obligated to ameliorate the circumstances of those with disabilities. Engelhardt, however, argues that (1) considerations of justice are not rooted in a common theory of morality and (2) one cannot defend a particular theory of morality through "secular moral reflection" without "begging the question, arguing in a circle, or engaging in an infinite regress." Engelhardt concludes by noting that if one is committed to establishing claim rights for the disabled then such rights must be secured through political compromise rather than moral argument.

Engelhardt is arguing for a very strong claim, namely that secular moral reasoning cannot be used to generate *any* claim rights for those with disabilities. The soundness of Engelhardt's argument depends largely on whether the principle that "we must treat equals as equals" should be rejected, as most if not all disability rights claims are grounded in this principle. While it is true that there is disagreement about how this principle should be specified, it is hard to dismiss the principle itself on rational grounds. However, if such a principle cannot be rejected then this implies that (1) there is a strong possibility that persons with disabilities *do* have claim rights (though there may be some disagreement as to what such rights consist in), and (2) such rights, if they exist, are rooted in considerations of justice.

1.4 Personal Voices

While much philosophical thought tends to be couched in what may be termed the "objective perspective," it is also appropriate to consider what might be termed the "subjective perspective." That is, focusing on and analyzing the personal experiences of individuals may shed much light on what many consider to be perplexing philosophical issues. It might be argued that this is particularly true in the case of the philosophy of disability. Many of those who lack disabilities may struggle to understand what it is like to live with disabilities. Or even more importantly, they may fail to understand what virtues can be cultivated by helping such persons.

In "Neither Victims Nor Heroes: Reflections from a Polio Person," Jean B. Elshtain begins by recalling her experience as an 11-year-old child with polio, when she was featured in her local newspaper as a child who "had been crowned a polio hero, having 'defeated' a crippling disease." Using this personal experience as a backdrop, she goes on to describes some of the ways in which the "hero" designation can be counterproductive, even harmful, in the lives of those so designated—for example, the pressure to "triumph-over" the effects of polio by using crutches, braces, or even a wheelchair, as well as undergoing potentially painful surgeries and lengthy hospitalizations. This is especially true for those persons who, having previously overcome the effects of polio itself during childhood and having lived for many years relatively symptom-free, now find themselves experiencing the degenerative effects of post-polio syndrome later in life. For such individuals, post-polio syndrome is often experienced as a *loss*, not only of physiological function but also of one's "hero" status, accompanied by return to a state of "victimhood." At the end of the day, Elshtain argues, in view of the equally dilatory effects of victimization and "heroization," persons with post-polio should be viewed as "neither victims nor heroes."

Elshtain's essay helpfully illuminates some of the central issues involved in the theoretical debates surrounding disability. For example, Elshtain's' observation that society "'names' us and [affects] how we think of ourselves" raises important considerations relating to stigmatization of the disabled, and how such stigmatization (to the extent that it exists) affects the self-conceptions of those with disabilities. Furthermore, her discussion of the "iconography of bathos," according to which those with disabilities (in this case, post-polios) are consistently and routinely characterized against a backdrop of victimization, brings to mind the frequent critiques in the disability literature of the "medical model" of disability; according to that critique, the medical model inevitably casts those with disabilities as being "victims" of a "personal tragedy," or as being "flawed" by virtue of having a physiological or mental deficit.

1.5 Conclusion

The various articles in this volume illustrate the fact that disability—whether considered from the perspective of conceptual analysis, ethics and public policy, or subjective experience—is a phenomenon that inevitably involves multiple frames of reference. This volume represents an attempt to capture the diversity of perspectives that one might adopt with respect to the complex philosophical questions surrounding disability. For that reason, we have not attempted to assemble a volume that promotes a single, overarching position on these issues. Instead, this volume is intended to be a *representative* sampling of a variety of approaches to the issues and questions under consideration. For similar reasons, we have not imposed upon our contributors any restrictions with respect to language or terminology; thus, the reader may notice differences in the language used to refer to persons with disabilities throughout the volume.

To conclude, the philosophy of disability is a multifaceted discipline. It is an area of inquiry in which metaphysics, phenomenology, and ethics intersect. It is our hope that through this work, the reader not only will better understand what disabilities are and how our society should respond to them, but also come to learn that disability is an aspect of the human condition which carries with it great meaning both for those with disabilities and for society at large.

Notes

1. The overview presented here is drawn primarily from Snyder (2006) and is, of necessity, cursory. For more comprehensive coverage of the matters discussed here, see Asch & Wasserman, 2006; Barnes & Mercer, 2003; Barnes, Oliver, & Barton, 2002; Burgdorf, 2006; Fleischer & Zames, 2001; Peters, 2006; Pfeiffer, 1994; Scotch, 1989; Shakespeare, 2006; Shakespeare et al., 2006; and Snyder & Mitchell, 2006.
2. According to Asch and Wasserman (2006, pp. 165–166), areas of agreement between the fields include similar concerns—e.g., concerns regarding "professional domination" by "experts" (physicians and other medical professionals), as well as their respective "demands for self-determination and autonomy"—and a parallel shift in focus to include "calls for sweeping societal change." Furthermore, both fields share similar emphases on patient autonomy, a skepticism regarding professional authority and the dangers of paternalism, and a strong support of "consumer rights."

References

Asch, A., & Wasserman, D. (2006). Bioethics. In G. L. Albrecht (Ed.), *Encyclopedia of disability* (pp. 165–171). Thousand Oaks, CA: Sage Publications, Inc.

Barnes, C., & Mercer, G. (2003). *Disability*. Cambridge, UK; Malden, MA: Polity Press; Blackwell Publishers.

Barnes, C., Oliver, M., & Barton, L. (2002). *Disability studies today*. Cambridge, UK; Malden, MA: Polity Press; Blackwell Publishers.

Buchanan, A., Brock, D. W., Daniels, N., & Wikler, D. (2000). *From chance to choice: Genetics and justice*. Cambridge, UK; New York, NY: Cambridge University Press.

Burgdorf, R. L., Jr. (2006). Americans with Disabilities Act of 1990 (US). In G. L. Albrecht (Ed.), *Encyclopedia of disability* (pp. 93–100). Thousand Oaks, CA: Sage Publications, Inc.

Charlton, J. I. (2000). *Nothing about us without us: Disability oppression and empowerment*. Berkeley, CA: University of California Press.

Fleischer, D. Z., & Zames, F. (2001). *The disability rights movement: From charity to confrontation*. Philadelphia, PA: Temple University Press.

Peters, S. J. (2006). Disability culture. In G. L. Albrecht (Ed.), *Encyclopedia of disability* (pp. 412–420). Thousand Oaks, CA: Sage Publications, Inc.

Pfeiffer, D. (1994). Eugenics and disability discrimination. *Disability & Society, 9*, 481–499.

Scotch, R. K. (1989). Politics and policy in the history of the disability rights movement. *The Milbank Quarterly, 67*, 380–400.

Shakespeare, T. (2006). *Disability rights and wrongs*. London and New York: Routledge.

Shakespeare, T., Bickenbach, J. E., Pfeiffer, D., & Watson, N. (2006). Models. In G. L. Albrecht (Ed.), *Encyclopedia of disability* (pp. 1101–1108). Thousand Oaks, CA: Sage Publications, Inc.

Snyder, S. L. (2006). Disability studies. In G. L. Albrecht (Ed.), *Encyclopedia of disability* (pp. 478–490). Thousand Oaks, CA: Sage Publications, Inc.

Snyder, S. L., & Mitchell, D. T. (2006). *Cultural locations of disability*. Chicago: University of Chicago Press.

Part I
Concepts and Theories of Disability

Chapter 2
An Essay on Modeling: The Social Model of Disability

Anita Silvers

2.1 Introduction

From its first exposition almost half a century ago, the social model of disability has been aimed at altering both theory and practice, bringing about profound changes in people's understanding of disability, and in the daily lives of disabled people as well. The social model's foil, and on some accounts its antithesis, is the medical model of disability. Both models treat disability as a locus of difficulties. While the medical model takes disability to be a problem requiring medical intervention—and as both the prerogative and the responsibility of medical professionals to fix—the social model understands disability as a political problem calling for corrective action by citizen activists who alter other people's attitudes and reform the practices of the state. These two conceptualizations of disability have been treated as competitors, as if one must prevail over and eradicate the other in thinking about who disabled people are and what should be said and done in regard to them.

Not long ago, the social model enjoyed almost unwavering allegiance from both disability activists and disability studies scholars. Now, however, fault lines in the disability community's fealty to the social model have appeared. Some disability studies scholars have launched criticisms of the social model, or more precisely, of claims they believe to be constitutive elements or entailments of the social model. These criticisms are advanced as being in the interest of revising disability theory to more accurately reflect disabled people's experiences, priorities, and needs.

The social model stands accused in some quarters of misrepresenting disabled people by abridging who they are, or of even more malignant distortions such as promoting values that exclude people with certain kinds of physical or cognitive limitations. These complaints are connected, in that the former criticizes the social model for suppressing rather than showcasing disabled people's differences, especially dysfunctional ones, while the latter objects to advancing an ideal of independence that some disabled people's dysfunctions make unrealizable for them.

A. Silvers (✉)
Department of Philosophy, San Francisco State University, San Francisco, California, USA
e-mail: asilvers@sfsu.edu

D.C. Ralston, J. Ho (eds.), *Philosophical Reflections on Disability*, Philosophy and Medicine 104, DOI 10.1007/978-90-481-2477-0_2,
© Springer Science+Business Media B.V. 2010

Pursued within disability studies scholarship, these charges echo concerns adherents of the medical model bring, namely, that to ignore experiences of being weak, enervated, in pain and vulnerable in modeling disability is deceptive because these are the most salient experiences in most, or at least in many, disabled people's lives. Of course, almost all people, regardless of whether they are disabled, have occasion to learn how such experiences feel. Yet some or all of these feelings chronically pervade the lives of at least some people with disabilities to a degree so marked and therefore so different as to distinguish their embodied lives, discerned at both the sentient and social levels.

These distinctive marks come to inscribe or inflect (some) disabled individuals' embodied subjectivity. Within disability studies, critics of the social model argue that obtrusive experiences of this sort, that seem to reflect minds or bodies (or both) while inflecting individuals' awareness or consciousness of their minds or bodies (or both), shape virtually every disabled person's awareness. Different disabled people respond in different ways, of course, but such characterizations of disabled embodiment are taken by many disability studies scholars (and by medical-model-influenced scholarship as well) as indispensable to understanding disability.

For example, bioethicist Jackie Leach Scully insists that "the strong social model is just not that interested in the subjective experience of the impaired body, or its psychoemotional aspects, or the processes through which disability is constructed by cultural representations and language." (Scully, 2008, p. 27) Scully adds, "The marginalization of disabled people cannot be effectively tackled, either theoretically or politically, if the subjective experience of impairment is left out" (Scully, 2008, p. 29). Scully and other critics (for example, Crow, 1996, p. 210) fear that the social model acquiesces to the traditional Cartesian split between mind and body, artificially splits the personal and the political apart, and fails to acknowledge that embodied perception and cognition distances disabled people's experiences from those of people who do not have anomalous bodies (Scully, 2008, pp. 28–29). Nevertheless, this line of criticism does not discount the social dimension of disability, and, despite returning focus to some of the matters that for the medical model are the essence of disability, need not adopt the values that motivate the medical model.

Parenthetically, by "strong social model," Scully explicitly has in mind the British "historical materialist" version of the social model (Scully, 2008, p. 28). In the US the social model also is invoked in the pursuit of political and legal liberation to argue for the social contingency of limitations that have been assumed in some quarters to characterize disability, but, *contra* Scully's account (Scully, 2008, p. 28) is decidedly concerned about the attitudinal and discursive manifestations of bias that energize and embed barriers (Silvers, 1998; Silvers & Stein, 2002; and see Areheart, 2008, for a recent example). This is not to say that the US version focuses on individual subjective states, any more than does its British forebear.

Ironically, at the same time (some) disability studies scholars are distancing themselves from the social model, medical professionals are drawing closer to it. An illustration is found in a recent Institute of Medicine (IOM) report. IOM followed up its 1991 and 1997 reports that designated disability as a pressing problem for public health by publishing a third report, in 2007, on *The Future of Disability in America*. Not unexpectedly, the IOM report is suffused with ideas and values associated with the medical model. For example, extolling the effectiveness of medical technology to prevent or remedy disability, the report applauds the reduction of activity-limiting biological dysfunction in older adults over the last two decades but warns that increases in physical inactivity, diabetes and obesity in the same time frame place younger and middle-aged adults at growing risk of disability.

Of course, such sentiments lie uneasily, to say the least, with the social model's commitment to altering social arrangements to make them more welcoming to biologically anomalous people, and the social model's opposition to altering biological individuals to prevent or fix their anomalies. Nevertheless, by no means is *The Future of Disability in America* an exercise in application of the medical model, for IOM announced the report's findings with words that appear to embrace the social model:

> Since IOM's previous reports in 1991 and 1997 that highlighted disability as a pressing public health issue, there has been growing recognition that disability is not inherent in individuals, but rather is the result of interactions between people and their physical and social environments. Many aspects of the environment contribute to limitations associated with disability—for example, inaccessible transportation systems and workplaces, restrictive health insurance policies, and telecommunications and computer technologies that do not consider people with vision, hearing, or other disabilities (Press release; see National Academies, 2007).

A further endorsement of the social model is added by Alan M. Jette, Ph.D., P.T., M.P.H., chair of the IOM's committee that produced the 2007 report: "Increasingly, scientific evidence reveals that disability results, in large part, from actions society and individuals take" (quoted in press release; see National Academies, 2007).

Which model of disability—medical or social—should shape thinking about disability's future in America? IOM's committee apparently saw no need to choose, an approach that may strike theoreticians immured in the debate between the models as question begging, while greeted by people to whom the debate has seemed peripheral as inspired. The discussion that follows here aims to illuminate the nature and purpose of creating models of disability, in order to see how the medical and social models of disability stand in relation to each other and whether there can be theoretical frameworks in which they coexist.

2.2 Models of Disability

Jette's mediating invocation of scientific confirmation as the basis for adopting the social model misunderstands what models are. Resolving the presumed conflict between the medical and social models is especially contentious because there is not nor can there be such a thing as a social model of disability. This concession does not gain much ground toward resolving whatever is in contention between the two accounts, however, for by the same token there can be no medical model of disability. Neither the ostensible medical model nor the so-called social model actually models disability, nor could any other set of claims of a similar nature do so. Indeed, what an empirical representation of disability would be like is highly unclear.

A model is a standard, example, image, simplified representation, style, design, or pattern, often executed in miniature so that its components all are easy to discern. Neither the medical nor the social model presents a replica or representation of disability. Sometimes appeals to models of disability are meant to invoke a standard or paradigm for categorizing people as disabled for a particular purpose, such as to determine eligibility for social insurance scheme benefits or statutory protection against disability discrimination, or to determine ineligibility for social roles such as employment or responsibilities such as parenting. But if not intended to play such a gate-keeping function, what are the nature and the import of the claims that constitute the supposed models of disability and that now are widely believed to be in contention with each other?

Even if the sets of claims labeled as models of disability do not exemplify or otherwise represent disability as one would expect an illuminating model to do, they may serve other purposes for which we often turn to models. They can, for one thing, help determine what and consequently when disability is. That is, they can delineate a paradigm to which people can appeal in deciding who is disabled. They can, for another thing, help to explain why disability is. That is, having settled on an account of what disability is, these theories can contribute an account of how people come to be disabled.

Models of disability thus may be called upon to facilitate two different purposes, sometimes singularly but sometimes both at once. First, a model of disability may be used to characterize disability identity and sometimes also to determine who is eligible to assume this identity. Competing models of disability may propose quite different properties as the qualifying ones. For example, the medical and social models portray disability in very different ways, the former in terms of biological defect and the latter in terms of social victimization.

Second, a model of disability may be used to explain why individuals are disabled (or, more explicitly, why they have the limitations associated with disability). Competing models may propose quite different causal accounts, together with quite different proposals about how to intervene in the causal process. The medical and social models explain disability in different ways and call for different courses of action to address it. On the medical model, freeing individuals from biological dysfunction is the recommended approach to alleviate suffering from disability, while the social model proposes that freeing disabled people from stigmatization and exclusion offers the most effective relief from suffering.

Identifying the properties that make an individual one of the relevant kind is different from explaining how the individual came to have those properties. It follows that the plausibility and power of competing models of disability may diverge depending on whether their classificatory or their explanatory effectiveness is being assessed. Whether the "what" purpose or the "why" purpose of conceptualizing disability is more significant becomes crucial when we want to judge the social model and decide whether it is superior or inferior to the medical model. Weighing the relative importance of the classificatory and explanatory roles requires a clearer notion of the circumstances in which people invoke the concept of disability

Identifying and disentangling from one another whatever discursive roles the social model of disability is used to play provides a better fix on what the social model of disability actually is, and thereby on the criteria by which to judge its adequacy. What the collection of claims that has become known as the social model of disability is supposed to do, and how well it satisfies the purpose(s) the social model is called on to achieve, will illuminate the relationship of the social model to the medical model (and to other so-called models of disability such as the moral model, the functional limitation model, and the minority model). The relative importance of the different roles, and how well the competing candidates can fulfill each of them, may cast light on which model should prevail.

2.3 The Concept of Disability: Classification

The idea that individuals with physical, sensory or cognitive impairments all together form a class of "the disabled" is a twentieth century invention. For in earlier times, classification was in terms of physical, sensory or cognitive condition. Persons were described as crippled or deaf or blind or mad or feebleminded, but only during the first part of the century was the term "disabled" introduced to characterize and collectivize them.

Disability as a concept originates in the context of the law, where it usually signifies a statutory incapacity or lack of legal qualification to do something. Someone with a legal disability suffers from an atypical or unusual or remarkable limitation that is legally imposed on her social participation, or at least is explicitly endorsed by the law. For example, prior to universal suffrage women had a disability in respect to exercising the franchise compared to men: they quite simply could not do so. Often, but by no means always, the legal limitations that constitute statutory disabilities are imposed because of supposed physical or mental inadequacies. Nineteenth century women were disabled in most places from voting, owning or managing their own property, and exercising custody over their own children because they were stereotyped as mentally feeble and physically frail. Analogously, during more than half of the twentieth century in the U.S., deaf people and blind people were prohibited from holding civil service jobs because they were stereotyped as inferior workers.

In contemporary Western culture, to be disabled is to be disadvantaged regardless of how much success one achieves individually. This is the generic implication of applying the term "disability" and its cognates and translations to label the group of biologically anomalous people whom we think of as being disabled. The idea of "disability" now is associated with physical or mental differences that compromise people's liberty to achieve typical levels of success in one or more areas of social participation, whether the relevant activities are learning, communicating, mobilizing, communicating, being employed or some other important productive activity. The key phenomenon that informs this idea is the experience of disabled people's being more limited than other people in one or more seemingly important respects.

Models of disability are invoked to identify the relevant kinds of limitations and to explain why these limitations occur.

Today, being physically, perceptually or cognitively impaired is categorically identified with disability, so much so that persons in very different conditions, with very different degrees of personal physical, sensory or cognitive limitation, and with quite disparate levels of socially significant achievement, are all referred to as "disabled." Some social model adherents have wanted to draw a sharp line between impairment and disability, thinking of impairment as natural, because biological, fact, in contrast to disability, an artificial social classification. There is nothing about social model theory, however, that entails or otherwise calls for this dichotomization.

Indeed, impairment itself has no fixed standard. What counts as being physically, perceptually or cognitively impaired is relative to the powers and limitations that are taken to be typical either of the species or of those members of the species who belong to a particular society or a prominent social group. What counts as being physically, perceptually or cognitively disabled is relative to how unusual a limitation is presumed to be. Being unable to fly is a species-typical human limitation that no one supposes to be a disability, although an eagle with such a limitation would be considered to be impaired.

2.4 The Concept of Disability: Limitation

Contemporaries' descriptions of seventeenth and eighteenth century people do not apply our categories of impairment and disability, although they do sometimes refer in detail to biological peculiarities or ill health. Perhaps because they were so much more common, expected and accepted then than they are today, illnesses, injuries, and syndromes (and their sequelae) that now place people in the disability group used not to render individuals socially unfit or invisible, that is, excluded on the basis of their biological anomalies. Biological conditions that since the nineteenth century have been subjected to therapeutic intervention or eugenic control were in earlier times accepted as common features of ordinary life, rather than as exceptional limitations.

The example of Samuel Johnson, who was the subject of replete writing by his friends, is a well-known illustration. Dr. Johnson was blind in one eye, had limited vision in the other, was deaf in one ear, was badly pock-marked, picked compulsively at his skin, suffered from spasticity or palsy and later in life from severe arthritis, and seems sometimes to have been so depressed as to remain bedridden (Boswell, 1934–1950, I: 485; Bate, 1978, *passim*; Thrale, 1984, p. 5). Nevertheless, as literary scholar Lennard Davis observes, "his contemporaries refer to his disabilities only in a casual and literary manner—tending to see him as a brilliant man who had some oddities rather than a seriously disabled person" (Davis, 2002, p. 49, and pp. 47–66, *passim*).

The initial expansion of the usage of the term "disability" to refer to physical, sensory and mental limitations appeared to move it from the political to the biological realm. What prompted the move was need for a terminology to refer to groups of individuals who, despite very different kinds of limitations, had been made eligible by statute for various kinds of benefits, such as compensation because their limitations resulted from injury during military service or supplementary income because their limitations predicted or justified their exclusion from the workplace. This extended application of the term afforded physicians considerable involvement in a gate-keeping role, delineating eligibility criteria for public and private disability insurance and social welfare schemes (Bickenbach, 1993, pp. 75–76). Because such benefits compensate for work limitations, not for being ill or injured *per se,* physicians have been tasked not only with diagnosing medical conditions but also with interpreting how limiting such conditions will be in workplace situations, despite their lack of expertise about all the ways to accomplish different kinds of work.

Yet both the amount of litigation over who actually is disabled, and the rapid growth of U.S. disability rolls (and those of many other Western nations) during the 1980s (when many kinds of diminished or inappropriate performances were medicalized), suggest the indeterminateness of biological evidence of disability. Compromised competence resulting from a biological anomaly—not the biological anomaly per se—is supposed to occasion the attribution of disability. And correlations between medically designated pathologies and limitations in competence are by no means reliable or firm (Stone, 1984, pp. 116–117, 128). These considerations suggest why it is problematic to identify biological anomalies as disabilities.

Further, the influence of environment on achievement, and therefore on impediments to achievement, is well-known. The ability of blind and visually impaired individuals to access inscribed information is a striking illustration of how environment affects functional limitation. When DOS was the predominant computer operating system, many blind and visually impaired people, using devices for reading the screen text aloud, embarked on careers that depended on the use of computer applications. Computers opened new avenues of productivity for people who previously could not access inscribed information absent specialized translations of the material and their own skill in reading Braille. But when the Windows program was marketed so aggressively that it eventually superseded DOS, Microsoft, claiming that business necessity demanded secrecy, refused to provide the computer codes for Windows to the specialized companies that developed voice output software. More and more applications ran only under Windows, and consequently blind and visually impaired individuals found themselves unable to perform the computer tasks essential to their jobs.

The environment thus affects the extent to which unusually limited vision functionally reduces access to acquiring and conveying important information. Small changes in common practice can have widespread and rapid effects. This writer first used the example of the limiting impact of Microsoft's domination of the software market on blind and visually impaired people in a book published slightly more than a decade ago (Silvers, 1998, pp. 107–110). The prognosis at that time threatened an outcome for blind and visually impaired people similar to what the telephone had imposed on deaf and hearing impaired people a century earlier. Despite having been invented during a project to improve communication access for deaf people, the introduction of the telephone proved disastrous because individuals who could not use it became extraordinarily limited in regard to functional communication in the workplace.

Fortunately, the Windows story has had a diametrically different conclusion. A political environment beginning to be reshaped by the social model of disability exhibited heightened commitment to equal opportunity for disabled people. Consequently, the organized campaigns of disability activists to make Windows-based programs accessible eventually elicited positive response when proprietary codes were made available to designers of adaptive software. Today blind and visually impaired people (and learning disabled people as well) can use technology that reads visually displayed texts out loud to access much of the information that is transmitted in electronic form. Indeed, we are moving toward a time when it is more accurate to talk about people being hard copy disabled rather than text disabled, a less disadvantageous limitation.

There are more reasons to believe that people's medical conditions underdetermine their functional limitations. To illustrate, individuals with identical prelingual hearing losses vary markedly in their ability to understand (usually by lip-reading), and be understood, through speaking. In the same vein, individuals with apparently identical muscular or nerve impairments differ in whether they can grasp, lift, stand, walk. There is a vast range of print-intensive occupations in which some people with dyslexia succeed while others with the same diagnosis fail, and this was equally the case prior to the medicalization and treatment of learning disabilities. These examples suggest that the limitations of or constraints on acting freely that disability is presumed to impose call for more expansive or more nuanced explanation than medical diagnoses of biological conditions usually provide. The superficiality of the biological account of why disability impinges on people's liberty suggests that this kind of model will not prompt the most propitious strategies for making their lives better.

Modeling disability in biological terms appears unable to account sufficiently for differences in the freedom of disabled and nondisabled people but especially for the reasons why the opportunities accessible to the former group usually are so limited. All models of disability seek to elucidate why individuals categorized as disabled are, in one or another way, unusually limited, and also, by explaining the reason(s) for these limitations, to show how individuals categorized as disabled can be less limited. But the explanatory power of the biological model is confounded by a multitude of cases in which the markedly different levels of achievement (and different degrees of suffering, as well) of individuals with identical biological conditions appear attributable to differences in how the individuals are socially situated.

2.5 Limitation and Political Action

The medical model elucidates disabled people's limitations in terms of biological pathology. This explanation makes medical intervention the route through which to address disabled people's limitations. If medicine can make them like other people—that is, cure them—the physical and intellectual barriers they encounter and that limit them will be no greater than what keeps the majority of people from being fully free.

The medical model treats the built and arranged environment as an invariable to which humans have no choice but to adjust. But it clearly is human to manipulate and alter our environment. We, through our social processes, fashion the built environment, which can be hostile or welcoming depending on how inclusively thoughtful public standards are. We, or at least the preeminent ones among us, also influence the organization of the dominant cooperative scheme which structures communication, citizenship, reciprocal contributions through work and civic duties, allocation of resources, and the other transactional processes of our social environment.

While the medical model presumes that disabilities are, fundamentally, deficits of natural assets rather than of social assets, the social model presumes exactly the opposite. And if disability is due to the disadvantageous arrangement of social assets that should be equitably accessible to everyone alike, social reform is at least as appropriate a vehicle as personal restoration for remedying the disadvantage disability brings. This was the message that, half a century ago during the era in which great improvement was gained in civil rights, began to be circulated by disabled people themselves, especially those who wrote about disability.

In 1966, Paul Hunt published an essay called "A Critical Condition" in a volume called *Stigma: The Experience of Disability*. (U.S. sociologist Erving Goffman's *Stigma: Notes on the Management of Spoiled Identity*, which made some of the same points but in a more abstract way, had been published three years earlier.) The thesis of this essay is widely accepted as the precursor of the social model. Hunt, a Briton with muscular dystrophy, had resided in institutions since beginning to use a wheelchair at age thirteen. His essay articulated what was then a radical view, namely, that what most limited people with disabilities was their segregation and the resulting social isolation.

Hunt attributed their social disadvantage not to their biological conditions preventing them from executing valuable social roles, but instead to their being the victims of socially embedded caricaturing that dismissed them as unfortunate, useless, different, oppressed, and sick. In other words, he analyzed their disadvantage as being of the same kind as that imposed on the victims of race-based and sex-based discrimination. Goffman was mainly interested in describing the psychological and sociological compliance in which disabled people engaged to survive in society despite such prejudice. Hunt, on the other hand, crafted a compelling, nuanced call for disabled people themselves to hold a moral mirror up to the nondisabled majority as a step toward liberating themselves from that same oppression.

A decade later, the British Union of the Physically Impaired Against Segregation developed an account of disability derived from Hunt's theme. In 1978 UPIAS proclaimed disability to be the disadvantage or restriction of activity caused by a contemporary social organization which takes little or no account of people who have physical impairments and thus excludes them from participation in the mainstream of social activities (UPIAS, 1978; Finkelstein, 1980; see also Barton, 1989 and Bynoe, Oliver, and Barnes, 1991). As its name suggests, UPIAS members saw their situation as an analogue to that of people of color such as black Africans suffering apartheid in South Africa. At that time, in the United Kingdom, the United States, and elsewhere, disabled people were often kept in charitable or state-run institutions regardless of their age, denied basic freedom to travel, to acquire a public education, and to pursue other opportunities their fellow citizens enjoyed.

In 1983 a research perspective was added when sociologist Mike Oliver, also an individual with a disability, labeled the ideas that lay behind the UPIAS definition as the social model of disability (Oliver, 1983). The social model removed individuals with disabilities from the role of dependent patient and recast them as independent citizens with rights which, when acknowledged, should eliminate the social disadvantages that are attendant upon their being a minority. From a health care ethics perspective, this transformation lines up with the evolution from paternalistic decision-making by professionals for patients to autonomous decision-making by patients themselves that prevailed during the same decades at the end of the twentieth century. Understanding that the burdens of exclusion and discrimination they bear arise from the defects of a biased environment rather than from personal deficits prompts both personal and political progress for people with disabilities.

The limitations associated with disability often are imposed or exacerbated by alterable cultural artifacts and arrangements. The social model uses this insight to trace the source of disabled people's disadvantage to a hostile environment and treats the dysfunction attendant on impairment as in great part artificial and remediable, rather than thoroughly natural and immutable. Thus the social model transforms the notion of handicapping condition from a state of a minority of people that disadvantages them in society to a state of society that disadvantages a minority of people. Their environment is inimical to them because in respect to almost all social venues and institutions, people with disabilities are neither numerous nor noticeable.

There are several compelling reasons why the social model is embraced by disabled people, and especially by disability activists and disability studies scholars. For one thing, the social model accords with disabled people's own experiences of the different effects of accessible and inaccessible environments on their lives. Whether they can execute the activities needed for daily life with the nonchalance other people enjoy, or whether even the most ordinary endeavors require a struggle to achieve, often is a matter of whether the organized environment acknowledges or ignores their existence.

From the standpoint of persons mobilizing in wheelchairs, for example, disablement is predominantly experienced not as the absence of walking but as the absence of access to bathrooms, theaters, transportation, the workplace, medical services and educational programs—all those opportunities citizens who can walk are at liberty to use. If the majority of people, instead of just a few, wheeled rather than walked, graceful spiral ramps instead of jarringly angular staircases would connect lower to upper floors of buildings. The wheelchair–using person who, in the course of a day of fulfilling responsibilities, must absorb the disruption of dealing with an insurmountable flight of stairs or an uncut curb, finds daily life to have a completely different texture in an environment where s/he can concentrate on ordinary work or play as other people do free of fear of encountering mobility barriers.

Were the existence of vision-impaired individuals afforded greater regard, information would not be conveyed in a format accessible only to the sighted. Tactile and aural modes of recording and conveying information would be used as frequently as printed texts. Today, electronically-inscribed information usually can be fashioned for speech as well as text output—that is, made to be heard as well as seen. In the past decade, to take just one illustration, dedicated political work by the blind and visually impaired communities has persuaded banks across the U.S. to install talking ATM machines. The social model accords with such experiences of the difference between respectfully and negligently organized environments and their impact on how freely the minority of disabled people can participate in the activities of daily life.

Suppose most people were deaf. Closed-captioning would always have been open and would have been the standard for television manufacture in the U.S. long before July 1, 1993. The 1990 Television Decoder Circuitry Act established the principle that for-profit makers of components required for public broadcasting have a responsibility to include deaf and hard-of-hearing citizens in the market segment they serve. In this case also, political work aimed at securing organizational recognition vastly increased the liberty of deaf and hard of hearing people to access important components of the social environment.

Parenthetically, in a footnote to an essay entitled "What Good Is the Social Model of Disability?" legal scholar Adam Samaha takes this line of argument (Samaha, 2007), which I have called "historical counterfactualizing" (Silvers, 1998, pp. 74–75), to be a political argument, requiring justification for the claim that the interests of disabled people who form a majority would or should prevail. Disabled people could be more numerous than other kinds of people and yet be incapable or undeserving of having their needs take priority. But historical counterfactualization is a thought experiment, not a political argument. Imagining most people to be situated as people in wheelchairs are—i.e., unable to access buildings that must be entered by stairs—it is easy to see that designing entrances with stairs would be pointless, as the resulting buildings would remain mostly empty. Historical counterfactualizing simply tests various exclusionary social arrangements to see whether something more than convenience for the majority accounts for those arrangements.

People with schizophrenia now are thought to be individuals with physiological impairments "that make them especially vulnerable to emotional stress... often from dealing with other people (which) can . . . spiral down into psychosis" (Grady, 1998, p. B17). Were the concern to maintain such individuals' productivity paramount, we would promote practices to reduce anxiety in interpersonal transactions instead of accepting—indeed, even admiring—the behavior of those who place stress on others while avoiding it themselves. Promoting supportive ways of relating to each other would acknowledge this dimension of neurological difference among people and make cooperative activities more accessible to more people by freeing those impeded by their reactions to adversarial practices to participate.

In sum, by explaining disabled people's limitations in terms of conditions that are subject to political action, the social model has empowered disabled people to achieve more freedom of social participation. A powerful reason for embracing it, therefore, is the proof provided by the improvements political action informed by the social model have made in shrinking the limitations disabled people experience in their lives. And as far as this reason goes, any incompatibility between the social and medical models is merely contingent and strategic. For neither model denies that both biological and social conditions contribute to disablement.

The decision about whether it is preferable to attempt to adjust environments to individuals' differences, or instead to alter the individuals so they more closely approximate the typical person for whom constructed environments usually are made, is to some extent a practical one. But it also is a matter of who is valued, which leads back to the question of what disability is. The social model explicitly explains why people with certain biological conditions may have less liberty than others, and the explanation is confirmed by disability activists' success in gaining liberty by pursuing political action to remove barriers in the architectural and technological environment. But simply understanding that inaccessible environments keep disabled people far away, or locked away, prompts no political reform unless value is placed on promoting social and economic inclusiveness.

2.6 Limitation and Disability Identity

Does the social model also adequately identify what disability is in terms that both those who are disabled and those who are not can embrace?

The social model ushers in an account of what disability is that is at odds with basic conceptions of medical ideology. Physical, perceptual and mental anomalies are not treated as flaws but instead as neutral human variations. The social model strengthens disabled people's positive sense of their own identities by refusing to measure them on scales calibrated to the typical human. On a standard constructed to make such normality the measure of man (and of woman), as the medical model does, disability inevitably falls short and therefore falls into the realm of the pathological (in contrast to being normal) or aberrant or "special" (in contrast to being ordinary).

In contrast, the social model aspires to employ a neutral concept of disability (Silvers, 2003). The anomalies that count as pathological on the medical model are portrayed by the social model in neutral terms as human variations. Such variations sometimes mean that the individual can function, but in a different mode. Thus, people who cannot walk mobilize on wheels, people who cannot hear use their hands to talk, and people who hear voices when there is no other person there have various strategies, as well as pharmaceuticals, for not responding to the voices.

Whether such alternative modes of functioning are available, and whether they succeed, depends on whether the environment fits the individual's adaptive abilities. Sometimes only tolerance of a different way of doing things is needed. For example, including a skilled lip reader in a spoken conversation often calls for no more than other participants' courteously facing the hearing impaired individual when they speak and making sure that hand gestures or shadows do not obscure the view of their lips. Sometimes the alternative mode of functioning requires the products of highly technological environments. For example, people with missing limbs will mobilize and manipulate successfully in an environment that can supply bionic prosthetics, and people with hearing loss will understand broadcast communications successfully in an environment where captioning devices are in common use. But sometimes the alternative mode finds more welcome in a non-technological society. For example, mild mental retardation that is dysfunctional in a highly technological society often is not even noticed in a subsistence farming community. And the person who lip reads instead of hearing clearly will communicate more successfully in a business environment like that which existed before the invention of the telephone, as well as that which now exists after the invention of email, than when commercial communication was conducted mainly by phone. Sometimes what is most important is acknowledgement of a disabled person's potential functionality. For example, 40 years ago no one with Down Syndrome could read or drive because the mistaken belief that they could not do so meant that no one bothered to teach them how, but now many such individuals do so, following upon legal mandates that they be provided with an equitable public education.

Sometimes, of course, biological anomaly results in inescapable dysfunction, either because the individual has no alternative mode of functioning or because the environment is not accepting of one. This being so, does the social model's strategy of distancing disability from dysfunction suppress what is most salient about disability in these cases, namely, that it is regrettable, or a harm? Common sense may prompt us to deny that disability can be a neutral idea, for it seems obviously preferable not to be disabled. Consequently, it is tempting to dismiss social model advocates as being in denial.

Of course, we cannot infer from our sense of one condition's being less preferable than some others that it also is inherently bad. We often prefer someone else's condition to our own—someone richer, smarter, handsomer, or more generous than ourselves—without condemning our own state as bad. Indeed, being dissatisfied with ourselves just because there are others who seem to us more fortunate is a recipe for unhappiness. So the fact that not being disabled may be preferable to being disabled does not entail that the state of being disabled is bad.

The social model counsels the acceptance of disability as being a natural state of some people, just as having a squarish shaped face, being five and a half feet tall, and remembering in pictures rather than in words are natural states of some people. Medical technology can be applied to alter each of these, but the cost, risk, and probability of failure in each case are disincentives to doing so. Even if disability is itself a neutral concept, however, any analysis of disability identity should address the relationship between disability and health.

Susan Wendell (2001) considers the ways in which people she calls the unhealthy disabled are different from the healthy disabled (Wendell, 2001, p. 19). The latter have relatively stable limitations, while fluctuations of limitation for those in the former category mean that their disabilities are not as easily understood. Sometimes they must endure other people's suspicion that they are not really disabled (Wendell, 2001, p. 21).

Like disability, illness should be a neutral category, according to Wendell. Illness is not evil in itself, but it causes suffering, which is evil (Wendell, 2001, p. 30). Some of the suffering that accompanies illness might be alleviated by improving social justice, but not all. Whereas healthy disabled people seek freedom from being confined to the "sick role," unhealthy disabled people may desire another kind of liberty, the freedom to dwell on their illnesses.

The feminist disability activist Jenny Morris describes what stands in the way of such freedom: "It is difficult... and dangerous because, to articulate any negative feelings about our experience of our bodies may be to play into the hands of those who feel that our lives are not worth living. We share a lot with other civil rights movements, but our form of oppression has a unique characteristic: it is not inherently distressing to be Black or a woman or gay, while it may be to experience an impairment.... But to deny the distressing nature of the body's experience of arthritis or epilepsy, for example, would be foolish" (Morris, 2001, p. 9; see also Morris, 1991, *passim*).

In Morris's experience, acknowledgement that one's health state is incurably defective is imprudent to express: "Sensory impairment, motor impairment, intellectual impairments are seen as things to be avoided at all costs. In the face of this prejudice it is very important to assert that anatomy is not destiny and that it is instead the disabling barriers 'out there' which determine the quality of our lives.... Indeed, I worry myself that if we do start talking about the negative aspects of living with impairment and illness, non-disabled people will turn around and say, 'there you are then, we always knew that your lives weren't worth living'" (Morris, 2001, p. 10).

"As long as non-disabled people retain the power to represent our reality," Morris says, "impairment will always mean at best a cause for treatment and cure, at worst a life not worth living.... It is this approach which leads to segregation and exclusion—and ultimately to the assumption that our lives are not worth living and that we would be better off dead, or not being born in the first place" (Morris, 2001, p. 3).

A somewhat less dramatic, but no less telling, argument along the same line may be found in analyses of how individuals with disabilities have been treated by U.S. courts in regard to the protection from employment discrimination promised by Title I of the Americans with Disabilities Act (ADA). Courts have tended to suppose that individuals who have been able to work successfully despite a biological or psychological impairment are insufficiently disabled to merit protection under the ADA, even when an employer's refusal to accommodate (for instance, to permit the employee continued access to an indoor parking lot with an elevator) prevents the individual from continuing in the job. On the other hand, courts (sometimes the same court) also have tended to hold that individuals who have not worked successfully where employers refuse accommodation are too disabled to be qualified for the job (Silvers, 1998; Pendo, 2002; Rovner, 2004; Areheart, 2008).

One conceptual response to these often repeated concerns has been to try to drive a theoretical wedge between impairments and people with impairments. The claim is that to abhor impairment is not to abhor people whose bodies or minds are impaired. But this strategy is far from reassuring, for impairments do not have abstract or disembodied existence, and negativity about an illness (or even a potential illness) easily transfers to negativity about people with that illness. For instance, in Chevron USA, Inc. v. Echazabal (2002) [00-1406], the U.S. Supreme Court ruled that an oil refinery could put an individual who long had tested positive for Hepatitis C out of work based on his potential for illness, even though he had been in the job and working successfully for over a quarter century without being symptomatic of either hepatitis or liver toxicity and with no unusual absences or other burdens placed on his employers or fellow workers (Silvers, 2005, 2007).

2.7 Can the Social and Medical Models Meet? Shake Hands? Connect?

Modeling disability in the ways both the medical and social models do mixes classificatory and explanatory discourse. This observation suggests where the models can be compatible, and also where they cannot. As we shall see, it is where explanation feeds action, and especially where explanatory theory translates into policy, that decisions granting exclusivity to one or another model have been, and perhaps must, be made.

Different systems of classification may focus on different features of their subjects, with the same individual being described in thoroughly different ways. Thus, for example, the same person may be accurately characterized as being both (socially) productive and (biologically) pathological. There is no logical conflict between these classifications, for we can conceive of people being socially productive despite being biologically ill or impaired, just as healthy but nonproductive people are conceivable. Nevertheless, challenging empirical disconnects and contentiousness between properties from different classificatory schemes may emerge or be fashioned. These sometimes occasion and sometimes even necessitate our making choices about which way(s) of classifying subjects to adopt.

Statistically, the correlation between the property of being socially productive and the property of being biologically anomalous may be weaker than holds for the productivity of biologically species-typical people, for example. That is to say, as a group, counting in all types of anomalies from the innocuous to the devastating, biologically anomalous people may be less productive along one, some, or all dimensions of productivity valued by a society than species-typical people are. When over-general and under-determined claims of this sort capture the public consciousness, policy tends to be adopted to drive the pathological and the productive further apart (than they actually are or need be), as for example, when the U.S. government adopted regulations that categorically excluded deaf and blind people from civil service jobs. Such background facts propel choices between modes of classification.

Historically, the contextual factors mentioned combined to place a premium on sorting people into or out of the workforce. Their salience promoted thinking about people solely or predominantly in terms of being normal or pathological. Classifying individuals in this way was a handy surrogate for designating them as "employable" or "nonemployable." Thus, because the social importance of work identities was so strong, the biological distinction between normal and pathological became a fundamental term in thinking about people in general.

And because the benefits of having a vigorous work identity were so conspicuous, explanations of how individuals came to fall into the pathological category issued in actions aimed at qualifying them for reclassification into the normal category, namely, at promoting programs for action aimed at reconfiguring individuals through the usual interventions into human biology—that is, at medical treatment through pharmaceuticals, prosthetics, surgery and rehabilitation. The medical model therefore is no representation of disability. Rather, it is a program for altering the numbers of people represented in the different categories of a classification scheme linked to the value of work.

To the extent that values other than work capability—for instance, liberty or security or happiness—evolve in preeminence, other systems of classification, together with programs enabling individuals to move from one category to another within them, will emerge. Liberty was the value that, at least initially, inspired the development of the social model, with its fundamental classification of people as institutionalized and therefore lacking liberty, or as living in the community and thereby free. In line with the analysis of the medical model advanced in the preceding paragraph, the social model thus can be understood as a program for altering the numbers of people in the different categories of a classification scheme linked to the value of liberty. Parenthetically, there is no more powerful example of a liberty driven disability program than the arguments for shifting the representation of disabled people from some legal categories to others advanced in the classic disability rights article by Jacobus tenBroek, "The Right to Live in the World: The Disabled in the Law of Torts" (tenBroek, 1966).

In a pluralistic society, we should expect that different models of disability will be appropriate to realize different values, and that these will be as compatible, or as antithetical, as the values they serve. Contentiousness between models generally can be traced to tension between values—for example, to the tenuous balance between security and liberty that must be maintained when a classification scheme incorporating categories of dependence and independence is invoked in modeling disability, as (for example) in some feminist care theory approaches to disability. As these are contests among values, we cannot expect science to confirm or disconfirm any model of disability, although weighing empirical evidence and attending to experience undoubtedly are important in considering the adequacy of any such model. Evidence and experience clearly contribute, along with values, to the process of setting a familiar model off to the side so as to think of disability in terms previously not conceived or supposed to be inconceivable. For a pluralistic society, many models of disability are better than one.

A caveat is called for here, however, for the corollary of the principle just articulated is not that any model will do, nor will any argument for appealing to or
rejecting a model do. To illustrate, the importance of health care for the population as a whole has been offered as a reason for maintaining the medical model
of disability against the social model. For the social model is feared to divert funding away from the health care system—for if disability is a property of environments
rather than of people populating those environments, then resources directed at these
people's health will be irrelevant to disability. If, as the social model suggests, people with biological anomalies, even with dysfunctional ones, can accept these and
flourish, if only their environment can be made more welcoming, the calculation
of benefit to risk in medical interventions to prevent or repair biological deficits
may change, with a concomitant diversion of resources away from healthcare, and
especially away from programs preventative of disability.

While there may be some merit to this worry from the general population's
point of view, models of disability should not be adopted to further the interests
of nondisabled people alone, or even predominantly. A basic question to answer
before relying on any classificatory or explanatory scheme is about what interests
the particular way of thinking serves. In their ways, and in their times, both the medical and the social models of disability have been welcomed as progressive. Each
now may create more concerns than benefits because the interests they have come
to serve have grown murky, or at least unclear.

The welcome news after half a century is that the social model's entrenchment
may be approaching the medical model's—especially if the medical community is
beginning to come on board, thereby achieving a practical balance between adjusting diverse people and uniform environments to one another. The challenging news
is that the philosophical struggle to align the fundamental values of which these two
models are expressions is nowhere near as close to achieving accommodation or
resolution. The most exciting news, however, is that people with disabilities appear
to become less and less marginalized when philosophical investigation of these values in the context of medical ethics and health care justice take place. Pursuing such
theoretical exploration rigorously but not dismissively helps purge both medical and
political understandings of disability of long ignored flaws caused by biased ideas
and oversimplified or simply false claims of facts.

References

Areheart, B. (2008). When disability isn't "just right": The entrenchment of the medical model and
     the Goldilocks dilemma. *Indiana Law Journal, 83*, 181–232.
Barton, L. (1989). *Disability and dependence.* Lewes: Faimer Press.
Bate, W. J. (1978). *Samuel Johnson.* New York: Harcourt Brace Jovanovich.
Bickenbach, J. (1993). *Physical disability and social policy.* Toronto: University of Toronto Press.
Boswell, J. (1934–1950). *Life of Samuel Johnson* (G. B. Hill, Eds.; Rev. L. F. Powell). Oxford:
     Clarendon Press.
Bynoe, I., Oliver, M., & Barnes, C. (Eds.). (1991). *Equal rights and disabled people: The case for
     a new law.* London: Institute of Public Policy Research.

Chevron USA, Inc. v. Echazabal. (2002). U.S. Supreme Court 00-1406.

Crow, L. (1996). Including all of our lives: Renewing the social model of disability. In J. Morris (Ed.), *Encounters with strangers: Feminism and disability* (pp. 206–222). London: Women's Press.

Davis, L. (2002). *Bending over backwards: Disability, dismodernism & other difficult positions.* New York: NYU Press.

Finkelstein, V. (1980). *Attitudes and disabled people.* Geneva: World Health Organisation.

Goffman, E. (1963). *Stigma: Notes on the management of spoiled identity.* New York: Simon & Schuster, Inc.

Grady, D. (1998). Studies of schizophrenia indicate psychotherapy. *NY Times* (Tuesday, January 20, B17).

Hunt, P. (Ed.). (1966). *Stigma: The experience of disability.* London: Geoffrey Chapman.

Morris, J. (1991). *Pride against prejudice: Transforming attitudes to disability.* London: The Women's Press.

Morris, J. (2001). Impairment and disability: constructing an ethics of care that promotes human rights. *Hypatia, 16*(4), 1–16.

National Academies. (2007). *Outdated policies are impediment for Americans with disabilities; Report recommends ways to remove barriers to care, assistive services.* Available at http://www8.nationalacademies.org/onpinews/newsitem.aspx?RecordID=04242007

Oliver, M. (1983). *Social work with disabled people.* London: Macmillan.

Pendo, E. (2002). Disability, doctors and dollars: Distinguishing the three faces of reasonable accommodations. *Disability, Doctors and Dollars, 35* U.C. Davis L. Rev. 1175, 1191.

Rovner, L. L. (2004). Disability, equality and identity. *Disability, Equality, and Identity, 55* ALA. L. Rev. 1043.

Samaha, A. (2007). *What good is the social model of disability?* 74 U CHI. L Rev. 1251, 1257. Working Paper 166, Public Law and Legal Theory Working Papers, University of Chicago Law School. Available at http://lawreview.uchicago.edu/issues/archive/v74/74_4/Samaha.pdf

Scully, J. L. (2008). *Disability bioethics: Moral bodies, moral difference.* Lanham, Maryland: Rowman and Littlefield.

Silvers, A. (1998). Formal justice. In A. Silvers, D. Wasserman, & M. Mahowald (Eds.), *Disability, difference, discrimination: Perspectives on justice in bioethics and public policy* (pp. 13–145). Lanham, Maryland: Rowman and Littlefield.

Silvers, A. (2003). On the possibility and desirability of constructing a neutral conception of disability. *Theoretical Medicine and Bioethics, 25*(6), 471–487.

Silvers, A. (2005, Winter). Protection or privilege? Reasonable accommodation, reverse discrimination, and the fair costs of repairing recognition for disabled people in the workforce. *The Journal of Gender, Race and Justice: A Journal of the University of Iowa College of Law, 34,* 561–594.

Silvers, A. (2007). Predictive genetic testing: Congruence of disability insurers' interests with the public interest. *Journal of Law, Medicine and Ethics, 35*(2), 52–58.

Silvers, A., & Stein, M. (2002). *Disability, equal protection, and the Supreme Court: Standing at the crossroads of progressive and retrogressive logic in constitutional classification,* 35 U. Mich. J.L. Reform 81.

tenBroek, J. (1966). The right to live in the world: The disabled in the law of torts. *California Law Review, 54* CAL. L. Rev. 841, 858).

Stone, D. (1984). *The disabled state.* Philadelphia: Temple University Press.

Thrale, H. (1984). *Dr. Johnson by Mrs. Thrale: The "Anecdotes" Mrs. Piozzi in their original form* (R. Ingrams, Ed.). London: Chatto and Windus.

Union of the Physically Impaired Against Segregation (UPIAS) and the Disability Alliance. (1978). *Fundamental principles of disability.* Available at http://www.leeds.ac.uk/disability-studies/archiveuk/UPIAS/fundamental%20principles.pdf

Wendell, S. (2001). Unhealthy disabled: Treating chronic illnesses as disabilities. *Hypatia, 16*(4), 17–33.

Chapter 3
Ability, Competence and Qualification: Fundamental Concepts in the Philosophy of Disability

Lennart Nordenfelt

3.1 Introduction

A large part of the population are every day absent from work and on sick leave because they are disabled (or claim that they are disabled) from doing their work, often but not always because of some illness, disorder or defect. This is, first, a great existential problem for the individuals who are affected and often for their loved ones. But it is also an economic problem for the people themselves and for society. To take an example: in 2003 on average 14% of the Swedish population were absent every day because of illness. The cost of sickness compensation that year amounted to no less than 110 billion Swedish crowns (approximately 17 billion US dollars). To this should be added the enormous cost for health services and economic losses due to reduced workplace productivity.

This crucial social fact calls for various kinds of research, and this in various disciplines. Apart from the obvious empirical questions which have to be answered there are fundamental theoretical questions suitable for a philosophical study. How should the concept of disability be determined, and what kind of disability in relation to work can entitle one to economic compensation? People can be disabled from doing things for a variety of reasons. They can have congenital defects that make them physically or intellectually unable to perform certain tasks; they can have problems in learning so that they cannot develop professional competencies and skills; they may lack stress tolerance or certain other kinds of tolerance in relation to the physical or social environment; they may lack the minimal self-confidence needed to appear at an ordinary workplace; they can have chronic diseases or they can be in acute crises due to the loss of loved ones, or due to other kinds of accidents or temporary illness. These are just a few examples.

Thus there are a number of potential causes of and reasons for a person's non-ability to take a job or fulfill his or her job duties. Although most of these causes and reasons are well-known, we lack a systematic analysis of them. There is a need

L. Nordenfelt (✉)
Department of Medical and Health Sciences, Linköping University, Linköping, Sweden
e-mail: lennart.nordenfelt@liu.se

D.C. Ralston, J. Ho (eds.), *Philosophical Reflections on Disability*, Philosophy and Medicine 104, DOI 10.1007/978-90-481-2477-0_3, to analyze how they are logically and empirically linked to each other. As a result of such an analysis we can distinguish between several kinds of conditions for action. A first well-known philosophical distinction is the one between ability, opportunity and will. All three factors must exist for an action to come about. The analysis in this chapter will concentrate on ability and its species. With regard to ability, crucial distinctions (to be explored below) include those between skill, competence and executive ability. A person may be perfectly skilled and in general competent to fulfill particular tasks but lack the executive ability to do so. In the context of work and work activities we must also take into account certain institutional conditions for action. People must have certain qualifications of a formal kind in order to get certain jobs (for instance a degree) and they must sometimes have a particular authority (for instance a position such as that of a judge or professor) in order to perform certain job-related actions.

This chapter is intended to be a contribution to the philosophy of disability. My approach, however, is unusual. I will analyze the concept of disability by way of considering its counterpart, viz. the concept of positive ability. I will analyze certain elements of ability, and I will distinguish between species of ability and consider their interrelations. I will also briefly consider some formal conditions the fulfillment of which is necessary for the performance of certain actions. As a result the concept of qualification will emerge. The entire discussion here will address the context of work and work activities. It is easily seen, however, that most of my observations are relevant to all contexts of human activity.

If one knows what are the sufficient conditions for the performance of an action, and what types of ability exist, one knows also indirectly what types of *non-ability* exist. The main result of the analysis in this chapter will be the categorization of types of ability and non-ability. This does not take us all the way to the notion of *disability*, as it is normally understood or as it could be understood, for instance in a medical context. Some but not all non-abilities would qualify as disabilities in an ordinary or medical context. Ignorance of a certain technique, for instance, may very well make a person unable to perform a specific task. This is, however, hardly a case of disability according to ordinary discourse. The ignorant person is looked upon as disabled only if he or she is also unable to *learn* to handle the technique in question and where this non-ability is independent of external factors.

This chapter places its emphasis on the basic task of identifying categories of ability and thereby categories of non-ability. The distinction between non-ability and disability will be briefly discussed toward the end of the chapter, but a deep analysis of this distinction requires a substantial study of its own.

3.2 Conditions for Action

There are a multitude of conditions that have to exist for an action to be realized. Apart from a bodily movement, a mental action, or an omission, there are many aspects of the external world that must be in order. The external world must provide the *opportunity* for the action to take place. And for the necessary bodily movement or mental action to occur the person must have the *ability* to perform it.

When a person has both ability and opportunity to perform an action, then we may say that there is a *practical possibility* of this person's performing the action in question. Practical possibility is the strongest form of ability. If, for instance, John has the practical possibility of driving his car and tries to do so, then John will succeed in driving his car. Trying could then be used as a test for practical possibility.

I shall first focus on the notions of ability and opportunity and their interrelations. Abilities and opportunities are concepts that indicate *dimension*. One can have more or less of an ability, and an opportunity can be more or less adequate. John can be a good driver or a bad driver, meaning that he has more or less of the ability to drive. A particular tennis court may provide a good opportunity to play tennis this year; last year, however, it was in poor condition and provided a bad opportunity.

Ability and opportunity are concepts that are logically interrelated in the following strong sense: when John is said to be able to drive his car, then this is so given a particular set of circumstances. John may be able to drive his car when the traffic is normal. He may, however, be unable to do so when there is a traffic jam. And, conversely, to take another example, when Sara is said to have an opportunity to play tennis, then this is so given a particular internal set-up in her case. The tennis court provides an opportunity to play tennis for Sara now that she is well-trained. Last year, however, when she knew nothing about tennis, the court would not give her any opportunity to play.

Thus there is no such thing as ability in isolation. And there is no such thing as an opportunity in isolation. A person's ability must be judged in the light of a certain set of circumstances. And a person's opportunity must be judged in the light of a certain set of conditions internal to his or her body or mind. But if ability and opportunity are in this way related to each other, what sense can we give to the idea of enabling a person to do something? In ordinary discourse we say that it is the duty of the health-care personnel to try to restore the person's ability to walk or to read. And normally we do not add anything about circumstances. Indeed, most of our ability talk is in absolute terms. We say about our fellow human beings that they can walk, drive cars, speak certain languages, etc. Strictly speaking, this must be elliptic talk. There cannot be any absolute abilities of these kinds. So what do we mean when we ascribe *abilities simpliciter* to people?

One can discern two important interpretations of this mode of speech. They are not rival candidates. I think it is clear that in some cases one interpretation is the true one. In other cases the other interpretation is probably correct. According to the first interpretation, a person is said to have ability, given that *standard circumstances* obtain. According to the second interpretation, he or she is said to have this ability, given that *reasonable circumstances* obtain. Let me comment on both alternatives.

In most cases when I claim that John is able to walk, I mean that this is so, given that there is nothing unusual that would prevent the execution of his action. The weather should not be extremely bad, the ground should not be extremely rough, and there should be no direct obstacles preventing him from walking. Given the way the world normally is—and in particular the way John's immediate surroundings normally are—John is able to walk.

Situations occur when the idea of standard circumstances fails to account for our talk about ability. Consider the following case of a schoolteacher in Iraq. He has been well-trained for his profession, he has a good talent for teaching and we would certainly describe him as a good teacher, i.e. able to teach young pupils. However, Iraq is a country that has for a long time been deprived of most reasonable opportunities regarding education. Most schools have been closed and there have been few if any possibilities of providing regular teaching. This has also meant that our schoolteacher has not been able to teach.

This situation of deprivation has for some time been the standard situation in Iraq. Hence the schoolteacher is unable, given standard circumstances, to teach. This is a strict application of the first interpretation. But certainly, we would say that he is capable of teaching. We must then have made a different presupposition. We must mean that he is able to teach, given *reasonable* circumstances. (For a more complete treatment of these concepts, see Nordenfelt, 1995, 2000, 2001.)

3.3 The Idea of a Standard Basic Competence

From birth we all obtain some biological and psychological preconditions for our future abilities. I call these *second-order abilities*. They are abilities, given standard or reasonable circumstances, to set oneself in a training program from which certain first-order abilities can follow. The latter notion can be defined in the following way: John has the first-order ability to perform an action A = def. John does A, if John wants to do A and there is an opportunity for John to do A. When we say of a person who is 18 that he or she can in the future work as a doctor we are talking about this person's second-order ability. He or she can study and train to become a doctor and subsequently perform the work of a doctor.

The training programs can be of several kinds. One may distinguish between, first, the standard life-training program that almost all infants and children go through and, second, the special training which is a prerequisite for certain professions or occupations. The former training includes learning to walk and learning to speak one's native language. It also involves basic school-training, learning to read and count and learning the elements of subjects such as history, biology and geography. All this also entails basic social training. I will refer to the results of this education and training as the *standard basic competence* of an ordinary adult in Western society.

It is true that not all people reach even the standard basic competence. Behind this competence already lie a number of physical and mental conditions, such as physical and mental health as well as a minimal degree of intelligence and perseverance.

For many purposes we can regard also the standard basic competence as a second-order ability for various occupations. It is from this competence that a person starts when he or she initiates a vocational or professional education. For certain menial jobs the standard basic competence is itself sufficient or almost sufficient for performing the job tasks. Examples are jobs as cleaners, doorkeepers or attendants. I emphasize here that I am talking about performing the job in a minimally acceptable way. There is a great distinction between being able to perform a job minimally and being able to do it well.

3.4 Occupational Competence and Skill

What does the special training program lead to? Ideally, it leads to occupational or professional know-how and skill. *Know-how* (a combination of relevant theoretical and practical knowledge) forms together with *skill* (the required dexterity) the person's *competence* with regard to a particular action. The trained person can, in principle, do what the job requires. I will add some further conditions later.

What does the training program contain? A few training programs, including the ones for becoming a police officer, a firefighter or an army officer, entail a lot of exercise where the basic physical and mental strength of the subject is developed. But all training programs provide some theoretical knowledge in a number of disciplines. The theoretical knowledge also entails a related problem-solving ability required by the profession. The electrician must not only know the theory of electricity but must also be able to understand what has gone wrong when the light goes out in a house. To this can be added all the practical skills that are necessary for performing his or her job well. The electrician must, for instance, be able—in a manual way—to handle the technical problems involved in repairing a stove or installing a refrigerator.

A good occupational training program should include much more. It should prepare the person for the various kinds of situations that he or she can meet which may not be saliently central to the profession. The program should consider the social milieu in which the person will be acting. It should stress the importance of communicative ability and of ethical competence. The person will, in almost all vocations and professions, meet with other people and deal with them, either in a curative, advisory, or perhaps commercial context. The program should prepare the student for various stressful situations and help him or her tolerate disruption, complaints and other kinds of frustration. The training program should also teach the student to constantly seek new knowledge and develop as a professional.

If the training program concerns an occupation that deals with the care of human beings or concerns leadership in an organization, certain requirements are deepened. Communicative ability will have a central role. To this can be added the ability to cooperate and establish deep contacts, to be able to support and comfort. As a leader one must also acquire strategic competence, the ability to make decisions and the ability to take responsibility.

In certain positions, including the leadership ones, a development of character may be required. A leader must have a certain amount of courage and ability to tolerate certain complicated and stressful social situations.

3.5 The Conditions for Pursuing Training Programs

Among the conditions for pursuing the vocational education or the training program we can discern the following:

- *Physiological or health conditions.* The trainee must for the most part fulfill minimal health conditions and must have a minimal physical and mental strength. In the case of the really demanding professions such as police officer, firefighter or army officer these requirements are substantial.
- *Opportunity.* The trainee must be given a fair opportunity to go through the program. The arrangements must be in order and the person must have sufficient time and peace to fulfill the requirements.
- *Intelligence and talent.* The trainee must be minimally gifted for pursuing the required tasks.
- *Capacity to cooperate.* Since most contemporary education and training programs include group projects, every participant must possess a minimal capacity to communicate and in general cooperate with the others.
- *Patience and perseverance.* The requirements in this regard are much more pronounced here than when we are talking about the conditions for the basic standard competence. The professional training program is more systematic and much more focused than the ordinary training for life. In many instances the tasks are quite difficult and the student may not be able to perform some of them in time.
- *Tolerance of stress and frustration.* The aforementioned difficulty introduces a new factor into the conditions. The subject must have a considerable stress tolerance and must be able to accept frustrations in the process.
- *Will and courage.* Again a crucial condition is that the subject has a standing intention to pursue the education or the training program. This intention must not be blocked by a fear on the part of the agent with regard to the continuation of the program. As soon as the will evaporates no action is pursued. The will thus plays a crucial role in the execution of all action. A close analysis of the concept of will lies, however, outside the scope of this presentation.

3.6 On Working Ability and its Conditions

3.6.1 Introduction

As I have said, the occupational education leads to (or is intended to lead to) occupational competence. This competence has two basic knowledge components: theoretical and practical knowledge. Theoretical knowledge is essentially knowledge in basic subjects such as mathematics, chemistry and biology but also applied sciences such as ship technology, caring science or museum technology. Practical knowledge is the knowledge of *what to do*, what concrete measures to take in certain crucial situations and how to perform these actions. I distinguish practical knowledge from skill in the following way. A person may have the practical knowledge to perform an action, but not the skill to do it. A pianist may have the practical knowledge to play a Beethoven sonata but at a particular moment not have the complete skill to do it, because he or she has not trained for the last couple of months. The pianist knows exactly "in the head" how the fingers should run and where to put an emphasis but will in practice not succeed in doing all this according to the formula. I will say that if a person has the theoretical knowledge, practical knowledge and skill required for a task then he or she has *competence* in the relevant respect.

I will in the following list some fundamental components of professional competence. It may be noted that this list is mixed in the following sense. It contains both species of competencies (or abilities to perform actions or complete tasks) and certain crucial conditions *behind* such competencies or abilities. Among such conditions are various kinds of knowledge, but also certain strengths and attitudes.

This distinction tends to be blurred by the fact that the term "ability" can be used in two ways. In the paradigm use of the general expressions "John is able to" or "John is capable of" these expressions are filled out with a locution referring to an intentional action on the part of John. We could say that John is able to teach French or John is able to build a house. Here teaching French and building a house are clear examples of intentional actions. Locutions exist, however, where ability can refer to mere dispositions. A child may be able to make a mess of a situation. A husband may be able to ruin his relationship to his wife. Here it need not be the case that the child and the husband actually intend to do so. The result may be an accident or at least an unintentional result.

It can be useful to make this distinction also in the work ability context. Normally the abilities and capacities cited in the following discussion refer directly to intentional actions. This holds for the ability to work hard or the ability to communicate. Ability in the mere dispositional sense appears, however, in the toleration and courage categories (see below). When a person tolerates the cold climate it is not a question (or not merely a question) of what the person intends to do. It is (at least partly) a question of his or her physiology and what this physiology permits. Similarly, the person who can tolerate or stand a noise may have the same intentions as one who cannot tolerate it. The ability or capacity in these cases then refers to properties or dispositions of the person other than his or her abilities to perform intentional actions.

One might then wonder if one should include these dispositions in the same category of work abilities. I choose to do this since they are certainly relevant in the context. For one thing, the intentional actions involved in work tasks are dependent on many of the dispositional capacities. Capacity to tolerate a certain amount of noise and stress is a crucial condition for one's ability to perform most actions involved in work. Thus the toleration category could be seen as including basic conditions for action on a par with physiological conditions, such as muscle strength and lung function. Similarly, courage could be looked upon as a kind of mental strength.

3.6.2 The Competencies

I will first highlight the category of technical competence, which is perhaps the one mostly referred to in applied documents concerning work competence and skill.

3.6.2.1 Technical Competence

Basic standard competence
Developed physical and mental strength
Developed intellectual capacity and talent
Theoretical knowledge, including problem solving-ability
Practical knowledge, including problem solving-ability
Adequate skill

I refer to the traits above as elements of technical competence. They are all to be seen as related to the specific vocation or profession. For instance, the theoretical knowledge and practical knowledge in question are such pieces of knowledge as are relevant for the work tasks of the individual in question. The list is intended to contain the minimal set of elements in a person's competence to perform "technical" tasks, where these tasks do not require continuous personal communication. Where communication and cooperation is necessary the list is inadequate. Since no job only entails purely technical tasks, but also involves some communication and cooperation, this list is always an inadequate list.

Observe here that I have summarized the technical competencies on a highly abstract level of description. It is clear that what I call developed physical and mental strength, intellectual capacity and talent, as well as adequate skill, incorporate a lot of specific abilities, some of which are specified (for instance) in the American *Handbook of Human Abilities* (Fleishman & Reilly, 1995). For instance, static strength—namely, the ability to use continuous muscle force in order to lift, push, pull or carry objects—is a crucial element in what I call physical strength. So is dynamic strength, namely the ability of the muscles to exert force repeatedly or continuously over a long period.

3.6.2.2 General Competence

Here follows a list of competencies which are not specific to any particular vocation but which are nevertheless relevant for almost all vocations:

Strategic ability, i.e. ability to plan one's work in a reasonable way
Adaptability to new situations
Ability to handle uncertainty
Ability to take decisions
Ability to take responsibility for the work done
Communicative ability

Ability to cooperate
Ability to assimilate new knowledge
Capacity to work hard

These items of general competence are certainly desirable for members of all occupations and professions. They are necessary for people who aspire to advanced positions in their professions. Caring professions and leadership positions require, however, a certain further set of abilities.

3.6.2.3 Personal Competence

Empathy, including

ability to establish deep personal contacts
ability to support and
ability to comfort other people.

Ethical competence, including knowledge about other people's rights
Ability to take decisions with regard to human beings
Ability to take responsibility for such decisions.

In what follows I will also list a number of abilities (or conditions of abilities) related to personality and character traits (of an ability type) which are necessary ingredients, at least to some minimal extent, in all occupations, but which are not commonly mentioned as included in professional competence. It is clear, though, that it would be desirable for all professional educations to highlight these as well as some of the personal competencies mentioned above.

3.6.2.4 Abilities Belonging to the Toleration Category

Ability to tolerate/withstand stress and heavy work-loads
Ability to tolerate the physical environment
Ability to tolerate the social environment

Specific items of these general categories are:

Ability to tolerate disruption
Ability to tolerate complaints
Ability to tolerate uncertainty
Ability to withstand frustration
Ability to tolerate criticism and opposition

Some of these abilities are typically included in the concepts of *patience and perseverance*, which belong to the category of virtues. I will return to this category below.

The toleration category (in particular when it comes to the non-intentional variants) is crucial in the health context. It covers a certain element of the person's fortitude, viz. the strength to avoid illness and damage. The stress-tolerant person is able to avoid the burnt-out syndrome better than the less stress-tolerant.

3.6.2.5 Abilities Belonging to the Courage Category

A requirement of courage—an important element of which is self-confidence—is salient, in particular with regard to crucial decisions to be taken by a leader, but it exists to some extent in all jobs. One has to have a minimal element of courage just to appear at the workplace and face one's workmates and other colleagues. Courage can be taught, but perhaps less so than other features of one's competence.

Having the courage to respond to criticism
Having the courage to counter dangers
Having the courage to make undesired decisions
Having the courage to oppose unsuitable proposals from the leadership

Courage is an interesting character trait. I have here, in a preliminary way, treated it as a species of ability. So does Per Bauhn in his *The Value of Courage* (2003). In general, he says, courage is the ability to confront fear. He distinguishes between two species of courage: first, *the courage of creativity*, which is the ability to confront the fear of failure (this ability being directed by the agent's will to achieve), and second, *the courage of conviction*, which is the ability to confront the fear of personal transience (for instance, being fired from one's job or being expelled from a community). This ability is directed by the agent's sense of moral responsibility.

What is lacking in this analysis is that courage must also contain an element of volition. The courageous person also has an attitude, an inclination. He or she has adopted a firm attitude towards a threat, and hence has a willingness to exercise this attitude. The person is willing to take a risk. This attitude may also be seen as a condition for the ability to confront fear.

Bauhn's two species of courage, the one of creativity and the one of conviction, are both relevant for the analysis of work ability. The worker must have some minimal courage of creativity in order to set about working at all. The worker must try to achieve and risk a failure. In some instances the risk of failure is great. A student working on a doctoral thesis takes a great risk of not achieving his or her goal. But this holds for all demanding jobs, such as doing professional sports, running a business and in general holding leadership positions. The courage of conviction is needed in situations of crisis, in particular when there is a situation of great danger and one has perhaps to act to save a life or when the worker chooses to stand up to the boss and expresses a radically divergent opinion, thereby risking his or her job. Admittedly this kind of courage is rarely executed and perhaps not very common. From a societal point of view I would, however, claim that such courage is a desirable property for any holder of a job. It is doubtful, however, whether all employers find it equally desirable.

A certain amount of courage, in particular the courage of creativity, is necessary for taking up and holding a job at all.

3.6.2.6 Other Virtues

Richard Smith (1987) has argued that in many professions, not least the profession of teacher, there is a requirement of certain virtues, such as wisdom, reliability, honesty and patience. His main point is that such virtues cannot be taught in a special skills training course. They belong to the deep personality of the person and can at most be the object of the long socialization process that takes place in the context of a good family.

How shall we look upon the basic virtues and their place in the structure of competencies? I think that most of such an analysis could parallel the one I have done with regard to courage. However, the virtues and other traits of character do not in general ontologically belong to the competencies as such. They are normally attitudes which could be seen as preconditions for certain competencies and abilities. Reliability and honesty are crucial virtues in relation to the whole work project. Reliability lies behind the fact that the work is done properly at the right time and with great care. Patience is a precondition for the proper continuation of the work in a context where there is much interruption and disturbance. Wisdom is that sophisticated virtue which only such persons have as are very highly experienced and have lived a long life. Wisdom presupposes some intelligence but is much more than intelligence. It involves a specific attitude towards mankind. It entails deep understanding of our human destiny and of how humans can react and develop. I think it also entails an attitude of benevolence. The wise teacher, for instance, abstains from reacting harshly to a student who has done a bad job when he or she understands that the circumstances behind the failure were outside the student's control.

3.7 The Notion of Qualification

A notion sometimes used in the context of occupations and professions is the one of qualification. How should it be differentiated from competence? In the literature the term "qualification" is used in different ways (see Ellström, 1997; Estes, 1974). I shall here interpret qualification as a concept covering certain formal conditions for professional performance. These conditions could be of various kinds, often of an institutional nature, for instance having a driver's license, having a clean conviction record, having citizenship and a military status. In the Catholic Church a woman is not qualified to be a priest. A man is normally not qualified to model women's clothes. A person under the age of 18 is unqualified for most occupations or professions regardless of his or her abilities. A certain certificate is needed for a person to qualify for many jobs. A person lacking a certificate but who is just as competent as the holder of a certificate is normally not qualified for the job. We might summarize the conditions mentioned here as *formal conditions* for entering an occupation or profession. It is typical that such formal conditions are crucial precisely at the stage of entrance to a job. At later stages they normally play a minor role.

In her study of service workers, Abiala (2000, p. 57) observes that, to a significant degree, many people in the service sector obtain their employment on the basis of qualifications other than competencies in the ordinary sense. In a questionnaire the people in the study group claimed that the following qualifications had been particularly important for their employment. The numbers refer to percentages of the study population (altogether 870 people):

Personality 88
Experience 65
Personal contacts 56
Sex 46
Age 39
Education 36
Looks 19

Race is clearly also in practice a factor of qualification in some communities. Research in the U.S. (Tilly & Moss, 1996) into the importance of so-called "soft skills" in the service economy has found that many employers perceive black men as inadequately qualified for the work required. According to these employers black men lack, in general, the desired attitude, behaviour and demeanor necessary for secure employment.

3.7.1 The Notion of Authority

A notion related to qualification is that of authority. Authority may be viewed as a special sort of ability, viz. a conventionally stipulated ability, which comes with the acquisition of a certain position. The authority grants that a desired outcome can be realized. Normally, only a police officer can arrest a person. Only a judge can sentence a criminal. Only a professor can examine Ph.D students. Here, as in the case of a certificate, there are rules stating when something is possible. When a person who is not a police officer grabs another person, the action can never count as an arrest. When a person other than a professor signs an examination document, this counts as a void action. Authority could be seen as a formal condition for the execution of certain tasks. If qualification in general is tied to entry into employment, authority could be seen as tied to the execution of a profession. In the following discussion, however, I will treat qualification and authority as being on a par with each other. I will use the term "qualification" as a general label for such formal conditions as are required for entering into employment and performing the tasks of a job. The category of qualification will together with competence and executive ability form a person's full ability.

3.8 Health and Executive Ability

To this shall now be added the crucial factor of *executive ability*, with regard to which health plays an essential role. One might wonder why executive ability or health is a factor to be added to the previously mentioned ones. I have already stated that the practical possibility for action is constituted by ability and opportunity. To what can health add? Is health a factor of opportunity? Indeed, health can be viewed as an "inner" opportunity. The state of the body gives the person an opportunity to act (in the case of health) or prevents the person from acting (in the case of ill health). This is, however, hardly a natural mode of speech. One may instead say that what I have so far summarized as factors of competence and qualification do not add up to a *complete* set of conditions for action, given standard or otherwise reasonable circumstances. It is not always sufficient for a person to be willing and completely competent or otherwise qualified in order for an action to come about. The person must also have a particular strength that is often lacking in a case of ill health. I will pursue this analysis further.

When the competent electrician catches the flu and does not go to work he or she does not, in general, lose the competence to do the work of an electrician. The basic competence is there all the time. What has happened, normally, is that the person has lost the strength to *execute* the competence during a short period. He or she has an aching body and has become very tired. The executive machinery is damaged.

Observe that ill health (in particular when it involves subjective suffering, such as pain, fatigue or anguish) also may strike at the person's *perception* of his or her work ability. In an informative study Schult, Söderback, and Jacobs (2000) studied two groups with comparable basic physical status. In one of the groups the participants abstained from work altogether mainly because of their subjective assessment of their status, whereas the other group attempted to do some work at least part-time. The second group scored better than the first one in most relevant respects, including subjective quality of life. The general mental attitude of the group that remained at home in fact reduced their work ability considerably, much beyond what could reasonably be expected given their physical status. The perception of one's work ability is thus in itself a considerable causal factor with regard to the ultimate work ability.

A disease or injury strikes primarily, I would argue, at executive ability. It does not typically affect overall competence. The latter is possible when it comes to certain serious and, in particular, enduring diseases. A neurological disease may reduce a person's intelligence as well as other elements of his or her personality. A chronic and serious disease of whatever kind will after a period reduce the person's talent for the job. He or she will inevitably forget both theoretical and practical elements involved in the competence. In addition, aspects of tolerance and courage will fade away. Thus ill health may in certain circumstances strike in a more basic way than just affecting the executive ability, i.e. it may affect the physiology of the person or the energy of the person. Ill health, in particular severe and chronic ill health, may certainly also affect a person's willingness to perform a job for a shorter or longer period.

3.9 A Summary of a Person's Internal Conditions for Work

I can now summarize the general conditions necessary for performing the tasks of a job. I have mentioned and discussed several items within the overall competence for the job. I have also indicated the necessity of the factor of volition, viz. the interest in the job and willingness to perform it in all its aspects. Moreover, I have added the notions of qualification and executive ability.

Let me then collect a person's inner conditions for work into the following seven major categories:

1. Overall competence for the job, including knowledge and skill. I have suggested that this competence should be divided into technical, general and personal competence. The overall competence also entails factors such as:
   a. Toleration of physical, psychological and social aspects of the job.
   b. Courage with regard to taking up the job and fulfilling the demanding tasks of the job.
   c. Virtues necessary for fulfilling the tasks of the job (examples: honesty, loyalty, perseverance and carefulness).
2. Qualifications for having and performing the job.
3. Executive ability to perform the job.
4. Willingness to take and perform the job.

3.10 From Non-ability to Disability

So far I have attempted to categorize some main elements in a person's full ability for work. I have hereby also implicitly identified the corresponding negative categories, viz. existing types of non-ability. However, as I indicated above, not all of these types of non-ability would qualify as types of disability in a medical or social context.

It is impossible within the limits of this chapter to do justice to the distinction between non-ability and disability. Moreover, it is not just one but several distinctions since there are different disability discourses. A social disability should be differentiated from a medical disability, to take one example. Let me, however, here make some observations and make a proposal which may function as a basis for a more thorough analysis of the notion of disability.

3.10.1 Disability as Related to Ill Health

One key category of disability is obviously related to ill health. The WHO's official classification of disabilities, the ICF (International Classification of Functioning, Disability and Health) from 2001, explicitly relates disabilities to a person's health problems. "The classification remains in the broad context of health and does not cover circumstances that are not health-related, such as those brought about by socioeconomic factors" (WHO, 2001, p. 8).

The ICF classifies both positive items called functionings and negative ones, viz. disabilities. The major subcategories of functioning are body functions and body structures, activities and participation. The major subcategories of disability are impairment (related to body function or body structure), activity limitations and participation restrictions. The subcategories are defined as follows. Body functions are primarily the physiological functions of body systems. It is notable, though, that this subcategory also includes mental functions. Body structures are anatomical parts of the body such as organs, limbs and their components. Impairments are problems in body function or structure constituting significant deviation or loss. Activity is the execution of a task or action by an individual. Activity limitations are difficulties an individual may have with regard to executing activities. Participation is involvement in a life situation. Participation restrictions are problems an individual may experience with regard to involvement in life situations.

Apart from mentioning the health context, the ICF does not further analyze the nature of the causes of disability. However, the predecessor of ICF, the ICIDH (International Classification of Impairments, Disabilities and Handicaps) from 1980 (revised 1999; see WHO, 1999) explicitly characterizes its categories as consequences of diseases. Diseases in their turn are to be found in the WHO's classification of diseases (WHO, 1992). Hence this notion of disability is closed within the health domain as defined by the WHO.

3.10.2 Disability as Related to Congenital Defects

This category of disabilities is still within the medical sphere. Indeed, the Congenital Anomalies form class XIV within the International Classification of Diseases and Related Conditions (ICD). However, the congenital defects or anomalies are worthy of special attention in this context. The main reason is that they are in ordinary discourse hardly ever referred to as diseases. Nor is the person with a congenital defect in general considered to be ill. Persons with scoliosis, a deformed arm or Down syndrome are neither regarded as suffering from a disease nor as being ill. On the other hand, all these persons are typically disabled persons.

3.10.3 Disability as Related to the Realization of Vital Goals

In my own analysis of the notion of disability (Nordenfelt, 2000) I have advocated a different characterization of disabilities. I agree with the WHO documents that not all non-abilities are disabilities. I also agree that there must be one or more factors intrinsic to the agent that are (together with the environment) causally responsible for the non-ability. I do not require, however, that the non-ability must be partially caused by a disease, impairment or defect. This moves me somewhat outside the strictly medical context. My criterion lies instead at the level of the goal of ability. I say that a disability is a non-ability to realize one or more of one's *vital goals* given standard or accepted circumstances (for the latter terms, see above). Thus a person is disabled with respect to a particular goal or action if, and only if, this person is unable to realize this goal or action given standard or accepted circumstances.

The notion of a vital goal requires some explication (for a more complete treatment, see Nordenfelt, 2001). I propose that a person's vital goals are the states of affairs which are necessary and jointly sufficient for his or her long-term minimal happiness. This idea could be rephrased informally in the following way. A vital goal is a state of affairs which is either a component of or otherwise necessary for the person's living a minimally decent life. This includes much more than mere survival. It includes life without much pain and suffering and it includes the realization of the most important projects of the person, such as having minimally decent accommodation, having a job and successfully raising children.

The resulting notion of disability is obviously individual-centred. A particular person's disability is crucially related to his or her particular vital goals. Does this not bring us too far from an established use of the term "disability"? One might wonder if it has intolerable consequences for the discourse on disability. Does it imply that there is no general way of describing disabilities?

I think there is little reason to fear such negative consequences. To start with, people's sets of vital goals are in practice not so particular. We all have, more or less for biological reasons, a number of vital goals in common. We all need to survive. Therefore most of us have to earn our living through work. As a further consequence we must all exist in a society with its typical institutions and norms. An important consequence of this is that it is a vital goal for all (or almost all) people to speak and otherwise use a language. This implies that impairments in the language "organs," including damage to the ear and the eye, will lead to disabilities for almost all people. Thus a nomenclature for such disabilities can have almost universal application.

On the other hand, it is true, in particular areas of disability, that there may be great variations among people. People's jobs, as we have seen, require different capacities; their leisure life may call for the use of varying abilities. This diversity, however, can hardly be a problem for international communication. It may be true that certain disabilities are rare, because of the rarity of the vital goals presupposed. The non-ability to take part in acrobatics is a disability only to circus-artists. The non-ability to hide a rabbit in one's coat is a disability only to a conjurer. But the rarity of an entity does not generally prevent us from having a nomenclature in other fields, let alone the science and practice of medicine. There are a great many rare diseases which are given names in the international scientific classifications. Admittedly, there is a logical difference between diseases and disabilities. And this is a difference that we must accept: in the case of disabilities two persons may be completely alike from a bodily point of view. They may both be unable to do F. It may, however, very well be the case that only one of them is disabled with respect to F. The person who is disabled has a vital goal that is such that doing F is necessary to achieve the goal.

3.10.4 On the Notion of a Social Disability

One might suspect that my vital goal notion of disability is still tied to a medical or at least a health context in general. Is there then no place for a notion of social disability distinct from health considerations?

I will mention three plausible interpretations of the notion of a social disability. Two of these, but not the third, are in line with my analysis. Let me start with the interpretation I wish to exclude. This is the case where the social disability is completely dependent on *external* circumstances. According to this a person is socially disabled when he or she is prevented from working or, in general, acting by external social obstacles. In the extreme case we have the imprisoned person who is prevented from moving outside the prison. Less extreme is the case where the lack of funds prevents a person from opening or keeping a business.

According to my theoretical analysis this is a case of lacking *opportunity*. The person is strictly speaking not disabled at all. Instead social circumstances are such that the agent does not have the practical possibility of doing what he or she wishes to do. (Observe that this is a consequence of my particular analysis. I certainly do not wish to legislate against a particular mode of speech.)

The other two interpretations are ones of *social causation*. The first case in fact brings us back to the medical paradigm. A person's social situation may be so stressful or even devastating that he or she gradually contracts serious illness. As a result the person becomes disabled in the basically medical sense. The second interpretation is similar from a logical point of view. Certain persons' social circumstances—for example, a loss of resources—may be such that without causing an obvious malady (a disease, impairment or defect) in the persons in question, those circumstances nevertheless change their personalities and inclinations to such an extent that there is no longer any question of their fully realizing their vital goals. Thus, these people may become disabled without there being a particular malady to diagnose. This would then be a paradigm case of social disability. Such disability could still be subsumed under my own general analysis of disability. The socially disabled person, in this sense, is unable to realize his or her vital goals given reasonable circumstances.[1]

Note

1. Portions of this work appeared, in slightly different form, in Nordenfelt (2008).

References

Abiala, K. (2000). *Säljande samspel: En sociologisk studie av privat servicearbete* [Selling interaction: A sociological study of private service work]. Uppsala: Almqvist & Wiksell International.

Bauhn, P. (2003). *The value of courage*. Lund: Nordic Academic Press.

Ellström, P.-E. (1997). The many meanings of occupational competence and qualification. In A. Brown (Ed.), *Promoting vocational education and training: European perspectives* (pp. 47–58). Hämeenlinna: Tampereen yliopiston opettajankoulutuslaitos.

Estes, J. (1974, Autumn). Welfare client employability: A model assessment system. *Public Welfare, 32*, 46–55.

Fleishman, E. A., & Reilly, M. E. (1995). *Handbook of human abilities: Definitions, measurements, and job task requirements*. Bethesda, MD: Management Research Institute, Inc.

Nordenfelt, L. (1995). *On the nature of health* (2nd, Rev. ed.). Dordrecht, The Netherlands: Kluwer.

Nordenfelt, L. (2000). *Action, ability and health: Essays in the philosophy of action and welfare*. Dordrecht, The Netherlands: Kluwer.

Nordenfelt, L. (2001). *Health, science, and ordinary language*. Amsterdam: Rodopi Publishers.

Nordenfelt, L. (2008). *The concept of work ability*. Brussels: PIE Peter Lang.

Schult, M.-L., Söderback, I., & Jacobs, K. (2000). Multidimensional aspects of work capability: A comparison between individuals who are working or not working because of chronic pain. *Work, 19*, 41–53.

Smith, R. (1987). Teaching on stilts: A critique of classroom skills. In M. Holt (Ed.), *Skills and vocationalism: The easy answer* (pp. 43–55). Milton Keynes, UK: Open University Press.

Tilly, C., & Moss, P. (1996). "Soft" skills and race: An investigation of black men's employment problems. *Work and Occupation, 10*, 252–276.

WHO (1992). ICD-10, *International Classification of Diseases and Related Conditions, tenth revision*. Geneva: World Health Organization.

WHO (1999). ICIDH, *International Classification of Impairments, Disabilities and Handicaps*, revised edition. (1st ed., 1980). Geneva: World Health Organization.

WHO (2001). ICF, *International Classification of Functioning, Disability and Health*. Geneva: World Health Organization.

Chapter 4
Disability and Medical Theory

Christopher Boorse

4.1 Introduction

This essay compares some contemporary ideas of disability to concepts in medicine, especially medical theory. Besides minor theses, I reach four major, not wholly original, conclusions. First, it does not seem that, in most usage, 'disability' has a clear enough meaning to determine how disability relates to medical status. Whether in common sense, ethics, or law, disability is a highly indeterminate concept. One reason is that the few paradigm cases at its core fall within many possible outer boundaries. Worse yet, even some core examples are disabilities in one context but not another. So, in practice, there is no one disability concept—or, if there is, it is ambiguous in including several variables fixed only by context. Second, all types of practical disability may be species of a value-free generic concept: organismic dysfunction, or gross impairment. But to call gross impairment disability will have consequences unattractive to many writers. Third, in two important contexts where usage is fairly clear by definition and example, disability currently bears no simple relation to the basic concept of medical theory, disorder or pathological condition. It appears that, for both the World Health Organization (WHO) and American disability-discrimination law, 'impairment' means nearly the same as "clinically evident pathological condition." But, in both editions of the WHO classification and in American law, an impairment is neither sufficient nor necessary for a disability. Fourth, in each practical context listed below, disability is clearly not a purely medical concept, theoretical or otherwise. Of two medically identical people, one can be disabled, the other not.

I first discuss why various fields of usage leave disability indeterminate, or, at best, contextually variable. After analyzing 'impairment', I discuss disability law and end by contrasting two approaches to disability advocacy.

C. Boorse (✉)
Department of Philosophy, University of Delaware, Newark, DE, USA
e-mail: cboorse@udel.edu

D.C. Ralston, J. Ho (eds.), *Philosophical Reflections on Disability*, Philosophy and Medicine 104, DOI 10.1007/978-90-481-2477-0_4,

4.2 What Kind of Concept is Disability?

4.2.1 Common Sense

One might suppose disability to be a common-sense concept. But one reason to doubt that the term has a clear ordinary meaning is the recency of its currently most popular use. Only in the last few decades has it become the usual umbrella term for paradigm cases like blindness, deafness, or paraplegia—replacing 'handicap', which, for obscure but partly ridiculous reasons,[1] was declared more demeaning. Indeed, when this essay was first written, the *Oxford English Dictionary* did not yet reflect this change. It offered only two general senses, neither of which fit the term's new role. The first sense is mere *inability*: "want of ability (to discharge any office or function); inability, incapacity, impotence," as in such archaic phrases as 'his disability to perform his promise.'[2] In this sense, of course, everyone suffers from innumerable disabilities, both physical and mental. Some all human beings share, like the inability to see ultraviolet light, swim the Pacific, mate with a dolphin, or mentally compute a thousand digits of π. Some are specific to individuals, like my own inability to speak Mandarin Chinese, play a Vivaldi piccolo concerto, run 100 meters in ten seconds, or bear and suckle a child. Presumably, none of these examples counts as a disability for recent disability law or literature. Yet the *OED*'s only other sense for the term was *legal incapacity*—a special restriction on the legal ability of an individual or group, such as children, women, Jews, felons, aliens, etc.[3] Clearly, this is not the meaning in current disability law or literature either.

Setting aside disabled vehicles, bombs, and nuclear reactors, I shall assume throughout that disabilities must be inabilities, either total or partial, of a whole person, or, perhaps, other organism. That is surely a minimal requirement—though too stringent, as we shall see, either for the U.S. Congress or for WHO. But if one thinks that common sense fixes which inabilities are disabilities, consider the following examples. No doubt blindness is a disability if anything is; but is color blindness? Red-green color-blindness? Myopia? If total deafness is a disability, what of partial deafness? Tone-deafness? Do the answers change if one is a painter or musician who loses a former ability to disease? It can be inconvenient to be left-handed, or at the 15th percentile of height; it can be life-threatening to be allergic to peanuts or wasp stings—but are these disabilities? Is a bout of flu a disability? If not, what of a relapsing infectious disease like malaria, or a regularly recurrent one like hay fever? Is controlled diabetes or hemophilia a disability? If one needs eyeglasses to read a newspaper, or a hearing aid to watch TV, is one disabled? Does it depend on whether reading is part of one's job? What of simple illiteracy? Is an illiterate adult disabled, or only if he cannot be taught to read? Is an English speaker disabled by moving to a Siberian town where everyone speaks Russian?[4] Why not, insofar as he resembles a stroke victim who gradually regains the power of speech?

As these cases suggest, intuition cannot answer many general questions about disability. How serious must an inability's effects be, or of what kind, for it to be a disability? Must it affect one's life expectancy, or one's ability to work, or to work at a specific job? Does whether an inability is a disability vary with environment? With age? Is a compensated defect still a disability? Can a disability be temporary, or must it be permanent, or at least recurrent? We must even ask whether everyone with a disability is disabled. Just as someone may have one peculiarity without our calling him a peculiar person, perhaps not everyone with a minor disability is disabled. Finally, can all organisms be disabled, or only human beings, or only some intermediate class? If ordinary usage leaves these questions unanswered, it seems equally unlikely to determine whether disabilities must be medical disorders, or, if so, what beyond medical abnormality is required.

One can deal with some of these issues by an adjectival strategy. That is, one can use 'disability' generically, for the widest plausible class, and add specifying adjectives at will: "total" vs. "partial," "temporary" vs. "permanent," "compensated" vs. "uncompensated," etc. As will appear, I take no position on whether biomedical science offers us any generic concept properly called disability. But the adjectival strategy cannot answer all our questions anyway. For example, is total color-blindness a total disability, a partial one, or none at all? Is sterility, or being a hemophilia carrier, a disability?

Further proof that 'disability' has no settled meaning is the wide variation in its usage among professionals in the field.

4.2.2 Usage By Specialists

That there is no standard meaning of 'disability' among health professionals is clear, first, from its differing usage by the three most influential "models" in the field: the two WHO frameworks and the various versions of Nagi's scheme.[5] Let us take these in historical order.

Beginning in the 1960s, working within Parsons' functionalist tradition in sociology, Saad Nagi distinguished *active pathology*, *impairment*, *functional limitation*, and *disability* (1965, pp. 101–103). While the first two are phenomena at the level of tissues, organs, or body systems, functional limitations are at the personal level and disability at the social level (1991, p. 322). Nagi described his four categories as follows.

> **active pathology** interruption of or interference with normal processes, and the simultaneous efforts of the organism to regain a normal state.

Nagi later noted that active pathology "may result from infection, trauma, metabolic imbalance, degenerative disease processes, or other etiology" (1991, p. 313).

> **impairment** anatomical, physiological, mental, or emotional abnormality or loss (1991, p. 322).

Impairments include: (a) all pathology; (b) residual loss or abnormality after active pathology ends; and (c) abnormalities (e.g., congenital) not associated with pathology (1991, p. 314).

> **functional limitation** limitation in performance at the level of the whole organism or person (1991, p. 322).

Impairment and functional limitation are abnormalities of function, but at different levels of organization: tissues, organs, or organ systems for impairment; the whole organism for functional limitation.

> **disability** limitation in performance of socially defined roles and tasks within a sociocultural and physical environment (ibid.).

Again, functional limitation and disability are both performance measures, but of "organismic" and "social" performance, respectively. Because of the social nature of disability, it is a "relational" concept, whereas the other three categories are pure "attributes" of the individual.[6]

To illustrate these categories, imagine that an office worker suffers a back wound that severs his spinal cord at his tenth thoracic vertebra. The wound is active pathology; the blockage of neural transmission to his lower spinal cord is impairment; his resulting inability to walk or run is functional limitation; and his inability to reach his job in a wheelchair-inaccessible office building, or to keep playing tennis with his wife, is disability.

The second most influential conceptual scheme appeared in 1980, when the World Health Organization (WHO), in its *International Classification of Impairments, Disabilities and Handicaps* (ICIDH), offered the following definitions:

> **impairment** any loss or abnormality of psychological, physiological or anatomical structure or function.
>
> **disability** any restriction or lack (resulting from an impairment) of ability to perform an activity in the manner or within the range considered normal for a human being.
>
> **handicap** a disadvantage for a given individual, resulting from an impairment or disability, that limits or prevents the fulfillment of a role that is normal (depending on age, sex, and social and cultural factors) for that individual (WHO, 1980, pp. 27–29).

ICIDH further describes impairment, disability, and handicap as disturbances at the organ level, personal level, and social levels, respectively (1980, pp. 14, 29).

Although the ICIDH and Nagi frameworks use 'disability' differently, this terminological difference masks obvious similarity in the two conceptual arrays. At first sight, ICIDH's "disability" looks like Nagi's "functional limitation,"[7] while ICIDH's "handicap" is Nagi's "disability." Actually, matters are not nearly so simple; there are many other contrasts between Nagi and ICIDH, and Nagi offers penetrating criticisms of its framework.[8] More important for present purposes are other criticisms of the ICIDH model by disability activists, especially those using a "social model"—criticisms that led WHO to a new conceptual scheme.

Although ICIDH acknowledged that social and cultural factors affect whether an impairment or disability causes a handicap, its scheme was unpopular with many disability activists. Rejecting the term 'handicap' altogether, they insist on using 'disability' in its stead to mean an effect of oppressive social barriers, in contrast to all intrinsic features of the individual. Early expressions of this semantic posture—which resembles a politicized version of Nagi's usage—came from the British group Union of the Physically Impaired Against Segregation (UPIAS) and from Disabled People's International (DPI). The UPIAS definitions were:

> **impairment** lacking part or all of a limb, or having a defective limb, organ or mechanism of the body.
>
> **disability** the disadvantage or restriction of activity caused by a contemporary social organization which takes no or little account of people who have physical impairments and thus excludes them from participation in the mainstream of social activities (UPIAS, 1976, pp. 3–4).

A few years later, the DPI offered the following:

> **impairment** the functional limitation within the individual caused by physical, mental or sensory impairment.
>
> **disability** the loss or limitation of opportunities to take part in the normal life of the community on an equal level with others due to physical and social barriers (DPI, 1982, cited in Barnes & Mercer, 2003, p. 66).

As a definition of 'impairment', of course, the DPI's first entry suffers from circularity. In any case, both the UPIAS and DPI formulations seem to wish to use 'impairment' to cover all of a person's intrinsic biomedical abnormalities—including, perhaps, the basic-activity restrictions which ICIDH calls "disabilities" and Nagi "functional limitations"—and to reserve 'disability' for socially caused disadvantages. That disability is caused by society, not by biomedical features of the individual, is the basic thesis of the "social model" of disability.

Yet a fourth professional usage of 'disability' occurs in WHO's revised document *International Classification of Functioning, Disability and Health* (ICF, sometimes called ICIDH-2), which aimed to meet criticism that its earlier scheme reflected an individual or medical model of disability. As Barnes and Mercer note, the new WHO classification

> sought to incorporate the 'medical' and 'social' models into a new 'biopsychosocial' approach. The overall result is a 'multi-purpose' classification system that retains the concept of impairment in body function and structure, but replaces 'disability' with activities, and 'handicap' with participation. In addition, ICIDH-2 assumes that functioning, activity and participation are influenced by a myriad of environmental factors, both material and social.[9]

ICF is explicit about these changes and their motivation.

> A variety of conceptual models has been proposed to understand and explain disability and functioning. These may be expressed in a dialectic of "medical model" versus "social model". The *medical model* views disability as a problem of the person, directly caused by disease, trauma or other health condition, which requires medical care provided in the form of individual treatment by professionals. Management of the disability is aimed at cure or the individual's adjustment and behaviour change. Medical care is viewed as the main issue, and at the political level the principal response is that of modifying or reforming health care policy. The *social model* of disability, on the other hand, sees the issue mainly as a socially created problem, and basically as a matter of the full integration of individuals into society. Disability is not an attribute of an individual, but rather a complex collection of

conditions, many of which are created by the social environment. Hence the management of the problem requires social action, and it is the collective responsibility of society at large to make the environmental modifications necessary for the full participation of people with disabilities in all areas of social life. The issue is therefore an attitudinal or ideological one requiring social change, which at the political level becomes a question of human rights. For this model disability is a political issue.
   ICF is based on an integration of these two opposing models (WHO, 2001, p. 20).

Unfortunately, the new terminology is much vaguer than the old.[10] Specifically, "activity limitations" are "difficulties an individual may have in executing activities" (an activity being a "task or action"), while "participation restrictions" are "problems an individual may experience in involvement in life situations" (2001, p. 10). So the original disability-handicap distinction, between failures at basic acts and their effects on social roles,[11] has disappeared. Nothing in the term 'task' or 'action' excludes a social role, and nothing in the term 'life situation' entails one. More importantly for present purposes, although 'handicap' is gone, 'disability' remains in ICF as an "umbrella term" for all three categories (2001, p. 3). Thus, it covers all impairments as well as limitations on activity or participation. I discuss impairment more fully below. But it seems that, in ICF, at least any clinically evident pathological condition counts as one. Thus, to take random examples, myopia (b21000), skin lesions (b810), hypertension (b4200), constipation (b5250), being overweight (b530), irregular menstruation (b6500), low sperm count (b6600), and baldness (b850) are all impairments, and therefore disabilities. ICF specifically notes that an impairment does not require any "capacity limitations"; e.g., "a disfigurement in leprosy may have no effect on a person's capacity" (2001, p. 19).
   Since mere local pathology, such as leprous lesions, is not an inability of the person, but at most of one of his parts, I suggest rejecting ICF's broad new usage of 'disability'. Even without it, the other frameworks show that there is no single meaning of 'disability' shared by writers, even professionals, using the term. What should we make of this? Perhaps there is, in fact, no common topic among disability writers. Alternatively, perhaps there is a common topic or topics, described in varying vocabulary. In particular, is there any generic concept of human activity limitation—whether called 'disability', 'handicap', 'functional limitation', or something else—underlying all practical disability judgments? If so, does it belong to biomedical science? Before discussing generic disability and some practical contexts, we must analyze 'impairment'.

4.2.3 The Medical Aspect of Disability: Impairment

Since, in disability literature, almost everyone uses 'impairment' to refer to the biomedical condition of disabled persons, one might expect impairment, at least, to be a medical concept of theory or practice. Whether it is or not, two possibilities arise: that all impairments are associated with[12] disabilities, or only some. At first sight, one expects *disability* to be logically parallel to *insanity* in criminal law: namely, to consist of a pathological condition severe enough to have certain morally and legally important effects. This expectation is often disappointed in disability literature.

Actually, the standard use of 'impairment' for biomedical aspects of a disability is curious, for there is little reason to think it a biomedical term at all, let alone a crucial one. Older medical dictionaries do not even list 'impairment', though they list 'disease' and 'pathological'.[13] On the other hand, recent editions often use the ICIDH definition, which comes, of course, from a reference work on disability, not medicine.[14] I do not know how 'impairment' got its current role as general biomedical term in disability literature. No doubt one motivation was that many disabilities—paralysis, blindness, missing limbs—are associated with static defects, not disease processes. ICF remarks: "Impairments are broader and more inclusive in scope than disorders or diseases; for example, the loss of a leg is an impairment of body structure, but not a disorder or a disease."[15] Actually, one can find examples of a very broad usage of 'disease' on which absence of a leg is called a disease.[16] I am not sure why WHO thinks that lacking a leg is not even a disorder.[17] It is certainly, in any case, a pathological condition, so it is unclear why any special nonmedical term had to be invented for disability literature.

In reality, the text of ICF, in its remarks and its examples, seems consistent with the thesis that *impairment* means *clinically evident pathological condition*. This is arguable on both historical and internal grounds.

As to history, the first edition (ICIDH) defined an impairment as "any loss or abnormality of psychological, physiological, or anatomical structure or function" (1980, p. 27). It also made two further, apparently restrictive, remarks. One is that "[i]mpairments represent disturbances at the organ level" (1980, p. 26), although later this was only said to be true "in principle" (1980, p. 47). Secondly, an impairment is called the "exteriorization of a pathological state" (1980, p. 47). This idea is one part of ICIDH's model of the (primary) disablement process (1980, p. 30):

What do the strange terms in italics mean? A pathological state is "exteriorized" simply when "someone becomes aware of" it—either the person who has it, or someone else, such as a health professional.[18] In other words, an impairment is a manifest or evident pathological state. This, however, is a different notion from organ dysfunction. Organ dysfunctions, such as a valvular defect, hearing loss, or early diabetes, may be subclinical, noticed by no one. Conversely, an evident pathological state need not cause organ dysfunction. A woman's breast may have a small, palpable cancer in it, yet perform perfectly its biological functions of lactation, attracting men, and giving its owner sexual pleasure. Up to a certain point, many organs, like heart or kidney, can functionally compensate for tissue disease by hypertrophy. So it is not true that clinically evident pathological states entail dysfunction at the organ level, as opposed to that of cells and tissues. Moreover, ICIDH itself says that an impairment can be "an anomaly, defect, or loss in a limb, organ, *tissue,*

or other structure of the body" (1980, pp. 27, 47; italics added). The mysterious phrase 'in principle' may be meant to smooth over this inconsistency.

The organ-level claim disappears in the second edition,[19] suggesting that ICF simply opts for the second of ICIDH's two definitions of impairment. Thus, though dropping the 'exteriorization' series of terms, ICF states that an impairment must be "detectable or noticeable by others or the person concerned by direct observation or by inference from observation" (2001, p. 12, n. 13). As to what is manifest, ICF defines impairments as "problems in body function or structure as a significant deviation or loss." This is just a less-technical version of the first edition's formula twice quoted above. It is time now to ask why all our writers so far define impairment disjunctively.[20]

Actually, "loss" of function can occur without disorder. A superior person's functioning might just drop to average without any pathological process, as by simple disuse. Suppose Martha, a famous pianist on a cruise, is shipwrecked and marooned five years on an island. When finally rescued, she has lost so much finger strength and dexterity that no one will hire her to play. Is she impaired? A textbook approved by the American Medical Association (AMA) gives a pianist example. It remarks that disability evaluation is not concerned with why someone cannot be a champion marathoner or concert pianist, but only with why someone can no longer be one (Demeter & Andersson, 2003, p. 4). In due course, this entry suggests that ability losses from nonmedical—presumably, nonpathological—causes can be impairments. But only medical impairments can be disabilities. So Martha is impaired, but not medically impaired, so not disabled.

> For the purposes of this book, an impairment is defined as deviation of an anatomic structure, physiologic function, intellectual capability, or emotional status from that which the individual possessed prior to an alteration in those structures or functions or from that expected from population norms. A disability is defined as a medical impairment that prevents an individual from performing specified intellectual, creative, adaptive, social, or physical functions (ibid.).

This textbook aims to be a companion to the AMA's *Guides*, which takes a similar view of impairment. The *Guides* define impairment as "a loss, loss of use, or derangement of any body part, organ system, or organ function," and mention that "medical" impairments can come from "illness or injury." Again a disjunctive view of impairment appears: "loss, loss of use, or derangement implies a change from a normal or 'preexisting' state" (Cocchiarella & Andersson, 2001, p. 2). Here 'normal' and 'preexisting' seem meant as possible alternatives. The authors again mention a superior person's ability loss—e.g., a gymnast who loses hypermobility, or a runner whose lung function is reduced merely to the population average. But the impairments described are due to injury, hence medical (Cocchiarella & Andersson, 2001, p. 4).

By contrast, ICF does not seem to use the category of nonmedical impairment at all. It says flatly that impairments involve "deviation from certain generally accepted population standards" (2001, p. 12), so medical abnormality is apparently required. To WHO, then, Martha is not impaired, nor even disabled in any sense if her rusty fingers are not a "health condition" (2001, p. 3). On this view, however, there is no need for a disjunctive definition of impairment. "Abnormality," in the relevant sense of subnormal function, is enough. Since congenital total nonfunction surely counts as an impairment, the "abnormality or loss" formula had to include it under abnormality. But then acquired total nonfunction must be abnormality as well. So there is no remaining independent conceptual role for "loss," if partial functional declines like Martha's, involving no medical abnormality, do not count. Medical abnormality becomes necessary and sufficient for impairment.

ICF further notes that "impairments are not the same as the underlying pathology, but are the manifestations of that pathology" (2001, p. 12). But if this merely means that impairment is the dysfunction caused by pathology in the medical sense of "lesion," then impairment coincides almost exactly with *pathological condition* or *disorder* as I have analyzed these terms.[21] The only obvious possible inconsistency between ICF's remarks and the pathological, as I define it, is over structural defects, like the missing leg.[22] I have held that purely structural abnormalities are not pathological. While some medical sources include them, they ought not do so, in my view, so as to preserve the simplicity of the concept of theoretical health.[23] A missing leg, of course, involves dysfunction too. Why, especially regarding disability, would structural without functional abnormality be of interest? Perhaps it is because both ICF and American law wish to include, under disabilities, social stigma evoked by mere physical abnormality. Even so, one can still identify *impairment* with *clinically evident pathological condition* if—like some medical works, but unlike me—one allows purely structural abnormalities to be pathological.

ICF's illustrative examples, supplementing my earlier list of minor impairments listed in the body of the book, also fit the thesis that 'impairment' means "clinically evident pathological condition." Annex 4 gives examples of seven conceptual categories, beginning with another case (besides leprosy) of "impairment leading to no limitation in capacity and no problem in performance."[24] A child is born with one missing fingernail, which "is an impairment of structure, but does not interfere with the function of the child's hand" or cause social stigma. In ICD-10—the corresponding WHO classification of medical disorders—a missing fingernail presumably falls under Q84.3 (anonychia) or, at worst, Q84.6 (other congenital malformations of nails). ICF's examples of "impairment leading to no limitation in capacity but to problems in performance" are (i) controlled juvenile diabetes, which causes problems in eating in a social setting where sugar intake is hard to limit, and (ii) facial vitiligo, which causes social stigma in a community that confuses it with contagious leprosy. Naturally, juvenile-onset diabetes (E10) and vitiligo (L80) are disorders in ICD-10.

In fact, most of ICF's examples involve clear ICD-10 disorders. The next three headings are cases of moderate mental retardation (F70 or F71), psychosis (various F-categories), and quadriplegia or inability to walk (G82). Only the two final examples challenge our thesis. A case of "suspected impairment leading to marked problems in performance without limitations in capacity" is a person who works with AIDS patients and is socially stigmatized because people "suspect he may have acquired the virus" (240). Now mere HIV positivity was not an impairment on page 19; nor is it apparently a disorder in ICD-10, but is rather in a special Z-category of miscellaneous "factors influencing health status and contact with health services."[25] So if "acquiring the virus" means becoming HIV-positive, this example seems to violate our thesis. However, it also contradicts ICF's own earlier text. Both problems would vanish if the example were, instead, suspicion of AIDS. Finally, an "impairment currently not classified in ICF leading to problems in performance" is a gene that increases breast-cancer risk in a woman whose mother died of the disease, and who is therefore denied health insurance. If the BRCA gene is an impairment, but one "currently not classified in ICF," then it can also be a pathological condition, but one currently not classified in ICD-10! We have, then, no clear counterexamples to our thesis about ICF usage of 'impairment.'[26]

But if an impairment is just a clinically evident pathological condition, then essentially everyone is impaired. This point does not even require allegedly universal, often subclinical, diseases such as atherosclerosis or tooth decay. Skin lesions alone, in their infinite variety, are visible on virtually anyone, any time. Almost everyone has a cut, bruise, insect sting, or blister, to name only a few possibilities from ICD-10's T14, not to mention warts (B07) and moles (D22). Whose skin is blemish-free? Only a fool would bet that a team of clinicians, backed by unlimited laboratory tests, could not truly assign him a single ICD-10 diagnosis. Rather, if 'normal' means "free of all pathological conditions," Edmond Murphy's quip was right: "a normal person is anyone who has not been sufficiently investigated" (1976, p. 123). This fact—which I am not aware is in dispute—has at least two implications for our topic. First, ICF's usage of 'disability', on which every impairment is a disability, is too deviant from other bodies of usage to be acceptable. It is false anyway that every cut or bruise lowers a person's ability, as opposed to the biological functions of a tiny area of his tissue. And typical activists, researchers, and writers of disability law assume disabled people to be a minority, not the whole population. As we saw, the UPIAS and DPI definitions describe disabled people as excluded from "the mainstream of social activities" and "the normal life of the community." It can hardly be that everyone is outside normal, mainstream social life, especially by virtue of a mosquito bite. Moreover, UPIAS described the disabled as "an oppressed group in society" (1976, p. 14), a view typical of the social model of disability. In a democracy, it is hard to see how the whole population can be an oppressed group. As for law, the Americans with Disabilities Act (ADA) calls disabled persons a "discrete and insular minority" and estimates their number at 43 million, or 17% of the 1990 U.S. population.[27] Similarly, the UK government has estimated that 20% of its population is covered by the ADA's British counterpart.[28] Thus, WHO's new usage, on which every impairment is a disability, is a poor one.

Second, one should ask what narrower scientific concepts are better candidates for generic disability. At least one medical dictionary requires an impairment to "interfere with normal activities" (*Mosby's Medical Dictionary*, 2002, p. 877)—activities, presumably, of the whole person. As we saw, ICIDH used 'disability' for this purpose, reserving 'handicap' for social disadvantages of impairment or disability. In whatever terminology, biomedical science certainly offers us a concept—indeed, various concepts—of a disorder affecting a whole organism.[29]

One is just biological dysfunction in an organism's gross capacity. A human being's, badger's, or sparrow's respiratory capacity may be species-subnormal, rather than only the gas exchange in one small area of lung's being so. A person's cardiac output, or his body's regulation of blood glucose or urea, may be subnormal, as opposed to someone with merely local disease of heart, pancreas, or kidney tissue. But such a concept barely differs from organ or organ-system dysfunction.[30] Alternatively, one can focus on basic activities of a normal human being—walking, lifting, eating, speaking, and so on.[31] Concepts like these two, which could be called "gross" or "organismic" dysfunction or impairment, fall squarely into biological science. As such, I have argued, they are objective and value-free (Boorse, 1976, 1977, 1997). The fact that an organism's capacities vary with its environment is no obstacle to the objectivity of biological dysfunction, as some disability radicals suppose. Not only do any species' organs have typical physiological capacity, the organisms pursue typical activities in the environments where they live. On nearly every analysis of biological function, organs have functions precisely by contributing to behaviors by which species members survive and reproduce in standard environments.[32] And human beings exhibit species-typical behavior just as do goats, crabs, spiders, or rosebushes.

That is not to say that everything physicians do in assessing "degree of permanent impairment" has a biological, or even a scientific, basis. The main body of the *AMA Guides* specifies, for each listed disorder, a "whole person impairment" (WPI) percentage. This it defines as the impairment's impact on a person's "overall ability to perform activities of daily living, *excluding work*" (*AMA Guides*, 2001, p. 4, Table 1-2). First, however, some of the activities that the *Guides* lists as commonly assessed are not species-typical—such as writing, typing, driving, flying—but merely important activities in societies like our own. Secondly, the *Guides'* definition is hard to reconcile with some of its listed percentages. Why is female infertility (2001, p. 169) a 30% WPI? The sole activities it blocks are conceiving and bearing one's own baby. It is consistent with every other aspect of parenthood and with all other activities. For that matter, why is total incontinence a 60% WPI, since it actually blocks almost none of the listed activities of daily living? One suspects that the *Guides* percentages are, in part, estimates of happiness or quality of life, not just of objective abilities. Fertility is very important to most young women, but that is not the same as its being required for most of their activities. Thirdly, to a person with two independent impairments of WPI percentages A and B, the *Guides* assigns a WPI of A + (1−A) B. So a totally incontinent infertile woman has WPI 72%; if also blind (85%), she reaches almost 96%—only 4% of her life remains. This formula succeeds in combining impairments without exceeding 1, but it may have no firm scientific basis, as the *Guides* admits.[33] In sum, the *Guides* may offer reasonably fair practical estimates of the disabling impact of pathology, especially for its main use, workmen's-compensation awards. But neither it nor textbooks suggest any reliable medical theory of degree of overall impairment. Whether biomedical science offers us an objective, value-free notion of degree of whole-person impairment may be the most interesting question in this area, and one on which my essay says nothing.

Regardless of degree, should we view gross, or organismic, impairment itself as disability—a generic disability concept anchored in biomedical science? Actually, this is not the view of any source discussed so far, though it is in the spirit of ICIDH's introduction.[34] One problem is that, like impairment, gross impairment may still be overbroad, counting too many people disabled. If normality is fixed by population statistics, no one specific functional deficit can afflict most people, but most can still have some gross dysfunction or other. A glance at prevalence data for some common conditions shows that this may well be so. The National Center for Health Statistics (NCHS, 2007) finds obesity affecting 34% of adults, while 31% of people over 18 have some joint pain, and 11% have had a severe headache or migraine in the past three months. More than a third of the population suffers from heartburn, with 10% afflicted daily, and 25–35% of American adults are myopic (Lange, 2006, pp. 1723, 2555). And the American Psychiatric Association finds premature ejaculation in 27% of men, with 10% having "erectile difficulties," while 25% of women have orgasm problems (APA, 2000, p. 538). Naturally, these patient groups overlap, but it would not be surprising if most American adults fall into at least one of them, besides the thousands of other possible diagnoses. And surely most adults middle-aged or older have some clinical disorder. Besides disorders above, hypertension afflicts 35% of adults aged 45–54, diabetes mellitus 11% of those 40–59 (NCHS, 2007), and osteoarthritis 30% of those 45–64 and 63–85% of those over 65 (Lange, 2006, p. 2701).

Thus, a gross-dysfunction test may well give most people a disability, and it all but certainly disables most people of at least middle age. Of course, this conclusion may be acceptable, even if one wants to restrict some benefit program to a minority. As noted below, most programs cover only disabilities of a certain type anyway. Or one might use the distinction suggested earlier, saying that all gross impairments are disabilities, but not all disabilities are disabling.

A second objection to calling gross impairment generic disability is that this allows organisms of any species to be disabled, from crickets to amoebas to slime molds to pachysandra. That sounds odd. Nevertheless, its oddness might be of a piece with that of calling plants and lower animals ill, which I have argued is of no theoretical significance (1997, pp. 11–12).

There may be, though, a third, converse, reason not to accept the equation "gross impairment = generic disability": to allow normal disability. Interestingly, three major sources discussed in this essay—ICIDH, ICF, and the ADA—all recognize disability without impairment.[35] Some of their cases involve past facts or mistaken beliefs about a person by others. But at least two normal phenomena that some sources call disability rest neither on the past nor on anyone's beliefs. One is a normal phase of life, like pregnancy, labor, or recovery, which limits abilities without pathology. Some writers put old age in this category. To medicine, it seems, a functional limitation typical of a person's age, like presbyopia or loss of cardiac muscle, is normal, not pathological. But some wish to call such a condition a disability.[36] Even needing glasses to read anything is now a serious handicap if glasses are unavailable. On this view, everyone who lives long enough is disabled; in the end, one escapes disability only by death.

A second phenomenon is mismatch between a normal variant and its environment. Such mismatches come in many types. One is a version of a normal species polymorphism occurring in a disadvantageous place. ICIDH's left-handedness example (1980, p. 167) is of this type, and perhaps also Nordenfelt's case (1987, p. 107) of lactose intolerance in Sweden. Although these are not species-abnormal traits, is that ethically or legally relevant if they cause local disadvantage? Another case is a person's normal, irreversible adaptation to an earlier environment. Organisms have normal capacities to adapt to a wide range of environments, but also normal limits to such flexibility. In particular, adaptability often falls sharply after a childhood critical period. Even nine months after adults move to a hot, humid climate, their sweat rate and water intake remain higher than natives'. Frisancho says: "it is doubtful that a low sweat rate can be acquired simply by long-term residence"; at the same time, tropical natives' advantage seems to be a "developmental response," not a genetic difference.[37] High-altitude adaptation is similar. Human beings who move to the mountains at age 2 achieve, as adults, about one-third higher aerobic capacity than those who move at age 16.[38] So, if a Vermont native takes a job in equatorial Brazil, or an adult Peruvian coast-dweller moves high into the Andes, their physical-labor capacity may be forever subnormal in their new environments. Yet their work handicaps are based on normal irreversible adaptations.[39] If there is a good reason why they must live in the new areas, it is, again, unclear why the fact that their conditions are not pathological is either morally or legally relevant to disability status.

In the generic sense, whether normal disabilities exist is open only to decision, not discovery. Medical usage, established for well over a century, settles whether pregnancy is pathological: it is not.[40] Nothing similar shows whether pregnancy is a disability, as opposed to one meriting coverage by a particular social program. Whether pregnancy is a disability in the generic sense depends, *inter alia*, on whether disabling conditions must be pathological, which usage does not settle.

4.2.4 Nonmedical Aspects of Disability: Ethics and Law

Let us now list some contexts of disability claims, to draw two conclusions. First is a nonlegal example: the "disabled list" of a sports team, familiarly abbreviated "DL" by sportswriters. Here, an athlete is disabled from a specific, context-definite activity: a game like baseball or football, or even a specific skill like sprinting. Note that a disabled athlete is not one who cannot play the game at all, but rather one who cannot play it safely, comfortably, at nearly his usual skill level, or some combination thereof. Note also that, though the sport may be the athlete's job, it need not be, since college teams also have a type of official disability.

Four more examples involve various areas of law. First is the humdrum example of handicap-parking license plates, which predate the ADA. Second and third are types of work disability. Many workers have disability insurance to protect their income. It may be private—a contractual employee benefit—or run by government, as are, in the U.S., state workmen's-compensation programs and the federal social-security disability program. Obviously, job-related disability insurance covers people who are disabled from work. But what work? In the private sector there are two main types of coverage, now called "own-occupation" and "any-occupation" plans. A plan may also combine the two, covering the worker for, say, one year of disablement from his own job but permanently for disablement from all jobs. As further complications, courts do not take "any occupation" literally, but count some jobs as irrelevant,[41] and many plans cover partial as well as total disability. Many countries also have special welfare benefits for those who have never been able to work, as in the U.S. social-security system's "supplemental security income" (SSI) program. Again, not all work counts the same; in 2007, disabled Americans could earn up to $900 per month without losing SSI benefits.

Finally, besides disability insurance for workers and disability welfare for non-workers, many governments ban various kinds of disability discrimination. In the U.S., under several federal statutes but most recently under the 1990 ADA, such discrimination is banned by employers of a certain size and in public services and accommodations. One expects the pool of people covered in their jobs by ADA-type laws to differ from the group covered by, e.g., disability insurance or welfare programs. To some extent, these measures aim at complementary groups: those who can work and those who can't. As discussed in Section 4.3.2, a disability for ADA employment-discrimination purposes can be in relation to a "major life activity," such as reproduction, unconnected with any job.

This list of types of disability claims—far from complete[42]—suggests two conclusions about disability. First, the concept is multidimensionally vague. The examples suggest that, in practice, disability judgments rest on a basic predicate with three or four argument places to be filled by context. To begin with, whenever anyone is called disabled, we must ask: "Disabled from what?" Context determines an *activity*, or range of activities, from which a person is disabled: a sport, a particular job, all jobs, the ADA's "major life activity," or something else. Since every conscious person can do something, it makes no sense to call a conscious person "disabled" *tout court*, without implicitly referring to some range of activities. Second, disability judgments are normally intended to have a practical *consequence*, prudential, moral, or legal: the person ought not be asked (or allowed) to play baseball today, or ought to get a cash income or reasonable accommodation in a workplace, public building, or parking lot. Whether or not such a consequence is part of the meaning of 'disability', writers tend to use 'disabled' as short for, as it were, 'disabled enough'. Practically, then, a disability is an inability of a certain type, severe enough to justify a certain consequence. Moreover, one must often specify the *environment* of the disability. That is why the ICF (2001, p. 15) distinguishes "capacity," a person's ability to do things in a standard environment, from "performance," his ability to do them in his current environment. So, in context,

x is disabled by impairment I

really means something like

Because x has impairment I, which significantly limits x in activities of type A in environments of type E, x deserves the consequence C.

My second conclusion is that, even given A, E, and C, the judgment whether x is disabled is not always purely medical.[43] Sometimes it is. If the spinal cord of a baseball pitcher covered by an "own-occupation" disability policy is severed at his third cervical vertebra, biomedical science alone proves him disabled, since a person with this injury cannot do the job at all—that is, pitch baseballs from the mound to the plate—without special equipment forbidden by the rules of the game.[44] But usually a disability judgment is not purely medical. Suppose the pitcher has only an inflamed shoulder, or a relay runner has a cold. Medical science may be able to assess probabilities of injury or of lower performance. But the judgment that the pitcher ought to rest his arm for a week, for his own sake or the team's future, or the judgment that the relay team ought not assume a certain risk of losing the race, is not a medical one, even of clinical medicine. Even apart from others' interests, a doctor can tell someone the risks of an activity, but not whether they are worth taking. Medically, one should not box or play professional football at all, but choose a less risky activity. Similar nonmedical issues arise in many job contexts, and even more clearly nonmedical ones in government programs. For example, under the ADA, a substantial limitation in a person's ability to work may be relative to "the geographical area to which the individual has reasonable access." So the same person may be disabled or not, depending on whether he lives in Philadelphia or Key West. Yet one cannot change one's medical condition by moving to a different city.[45]

Thus, IAEC contextual analysis highlights two points from my opening summary. First, in each context listed, of two medically identical people, only one may be disabled. The same elbow tendinitis may disable a pitcher but not a soccer player.[46] In other work outside sports, losing a ring finger may block a violinist's own occupation, but not an English professor's. Most Americans with Steven Hawking's pathology would claim Social Security benefits as totally unemployable, not work as leading physicists. Identical paraplegia justifies a handicap-parking plate for someone who has a car registration, but not for someone who doesn't. And under the ADA, as noted below, whether an infertile woman is disabled may depend on how much she wants babies. Second, a paradigm of disability in one context may not be a disability at all in another. Blindness and psychosis are core cases of disability in many contexts, but neither justifies special parking rights: schizophrenics can park anywhere, while blind people cannot park at all. And, obviously, a blind professor, on getting tenure, cannot stop working and claim 80% salary for life for his disability. Yet he is a paradigm of someone covered by the ADA or by SSI. In these and other ways, disability judgments vary with contextual factors independent of medicine.

4.3 American Law on Disability Discrimination

4.3.1 Can an Antidiscrimination Goal Define Disability?

In the background of much recent literature, one senses an assumption that *disability* means the kind of inability that is, or, at any rate, ought to be, covered by antidiscrimination laws like the ADA. This stance, no doubt, owes much to the belief that, under a proper law, ADA disabilities will be the largest class: anyone disabled in any other context will be disabled for the ADA too. Yet the goals of antidiscrimination law may fail to delimit disability in at least two ways. One is that laws like the ADA may be unjustified. Another is that, even if such laws are justified, their goals may define no unique protected class.

First, if disability is what the ADA should cover, then, if the ADA is unjustified, disabilities do not exist. And the ADA might be unjustified either because civil-rights law in general is unjustified, or because disability fails to fit its paradigm.

Basic statutory antidiscrimination law, beginning with the U.S. 1964 Civil Rights Act (CRA), enjoys the strong support of a huge majority both of Americans and of legal commentators. Still, it has its critics. Actually, the CRA's ban on private discrimination faces the same kinds of objections—libertarian, constitutional, and utilitarian—as in the standard liberal critique of drug laws. First, from political philosophy, is the libertarian objection: any legitimate government must respect a citizen's right to dispose of his own assets as he pleases, so long as he does not harm others in so doing. Regarding private discrimination, although groups like the American Civil Liberties Union have not so seen the matter, the true civil liberty is arguably to discriminate however one pleases—not an imaginary right to force other people into unwilling transactions. Second, from law, is the constitutional objection: the U.S. government, in particular, some argue, has no legal power to ban private discrimination, even under such pretexts as the commerce clause.[47] Last, from ethics, is the utilitarian objection: the actual good effects of the law are few and small, while its bad effects are many and grievous.[48] If, for any of these reasons, the private-discrimination provisions of the CRA are unjustified, and if disabilities are inabilities that should enjoy similar protection against private discrimination, then there are no disabilities.

Even if basic antidiscrimination law for categories such as race, national origin, religion, and sex is justified, disability may not resemble them enough to justify the ADA and its ilk. To inherit presumed justification from the CRA, the ADA must be a natural extension of it, rather than, say, a monstrous perversion. Yet textbook writers grant immediate moral differences between disability and other categories of discrimination. To begin with, disability is often rationally relevant (Tucker & Milani, 2004, p. 2; Zimmer, Sullivan, & White, 2008, p. 489). Such moral differences, in turn, lead to many contrasts between U.S. disability-discrimination law and other kinds. Unlike the CRA, the ADA gives employers a positive, often costly, duty to workers: reasonable accommodation. Employers' duty to spend money accommodating disabled workers—a duty with no counterpart in other kinds of discrimination—evokes at least two special criticisms of the ADA. First is the incurable vagueness of 'reasonable' (or of 'undue hardship'), given that the employer's economic rationality is not the test. Secondly, on any ethics that either denies any general duty to aid needy strangers, or agrees with Thomson (1971) that only "small sacrifices" are required, what the ADA forbids employers to do is not morally wrong. That would mark a second major difference between the ADA and its model, the CRA, since the basic immorality of discrimination by race, national origin, religion, or sex was a premise of the civil-rights revolution.

Admittedly, the above criticisms mainly touch private discrimination. The ADA's requirements on government might still be justified, so that the class it properly protects could still delimit disability. Yet we seem to expect nothing similar in other fields of discrimination. We do not expect the CRA to tell us what a race, national origin, religion, or sex is. The law presupposes these categories; it is not thought to create, or even to do much to delimit, them. By contrast, to apply the ADA, unlike other civil-rights statutes, one must judge who is in a protected class, since, unlike the CRA and other models, it does not protect everyone, but only a minority: disabled people.[49] As Colker notes, ADA plaintiffs often lose by a summary judgment that they do not have a disability in the first place.[50] Because so many ADA cases turn on this issue, they often show obscurity in its disability concept, especially in its key ideas of "substantial limitation" and "major life activity."

Finally, even if an ADA-type law is justified, case law reveals at least two distinct views of the goals of disability-discrimination law. At best, these views delimit two classes of disabilities. One view, enshrined at the beginning of the ADA and stressed by the *Sutton* majority, is that disabled people are a disadvantaged minority. The purpose of the ADA is to protect this minority from oppression by state and private action. Another view is that the ADA (or some revision) ought to protect a majority—everyone, perhaps, or at least everyone with a medical disorder—against, in Justice Stevens' words, "irrational and unjustified discrimination because of a characteristic that is beyond a person's control."[51] Given this contrast, there are at least two possibly justified ADA-type laws, generating a narrow or a broad class of disabled persons.[52] So one must be wary of assuming that an antidiscrimination goal, by itself, will ever make the disability concept clear.

4.3.2 Disability and Impairment in the ADA

Regardless of what, if anything, the ADA should say, let us now turn to the meaning of *disability* and *impairment* in this law as written and interpreted. As I said, one might expect disability, in discrimination law, to be a conjunctive medical-legal concept like criminal insanity. In Anglo-American law, all four major insanity tests use psychopathology as a necessary condition; the three still in use add further requirements for a mental disorder to be a criminal excuse.[53] Thus, by all current tests, insanity is mental disorder with enough further exculpatory properties to make it a complete defense to criminal guilt. Similarly, in disability-discrimination law, one might expect disabilities to be pathological conditions that meet tests of relevance or gravity. At first sight, this expectation is confirmed when writers describe disabilities as impairments with certain effects. However, contrary to expectation, it is not entirely clear that 'impairment' means "pathological condition" or even "clinically evident pathological condition."[54] The equation certainly does not hold in UK law. Moreover, even if it holds in US law, medical disorder is clearly still not a necessary condition of disability.

To take the latter point first, the ADA uses a strange three-prong structure, borrowed from an earlier federal law, the Rehabilitation Act, to define its coverage. As the ADA's title indicates, a protected person is an "individual with a disability." In turn:

> The term "disability" means, with respect to an individual –
> (A) a physical or mental impairment that substantially limits one or more of the major life activities of such individual;
> (B) a record of such an impairment; or
> (C) being regarded as having such an impairment.[55]

Thus, a disability is either a certain kind of present impairment—let us call it, as several texts do, a "substantially limiting impairment" (SLI)—or the fact of having once had an SLI in the past, or the fact of being mistakenly believed now to have a SLI. ("Mistakenly," because for true belief, (C) adds nothing beyond (A).) Awkwardly, the law's text uses no single word for SLI, the kind of impairment described by (A). But, obviously, 'disability' cannot be that word, else people relying on (B) and (C) would not be individuals with disabilities at all.

This bizarre usage is so counterintuitive that even experts, such as courts and legal treatises, fall into continual confusion or contradiction. As an example of confusion, Colker, after quoting the text above, writes:

> This definition has three prongs: (A) an actually disabled prong, (B) a record of disability prong, and (C) a regarded as disabled prong.... Even if one is not actually disabled at the time of discrimination (and thereby not covered by the first prong), one might be covered under the second or third prong.[56]

Here, obviously, Colker is using the term 'disabled' to mean "having an impairment of the kind described in (A)," a substantially limiting impairment (SLI). But it cannot mean that in the ADA. If the three clauses (A)-(C) define disability, then disabilities under clauses (B) and (C) are just as "actual" and "present" as disabilities under (A). Like Colker, a leading casebook states: "ADA coverage does not depend on establishing an actual, present disability" (Zimmer et al., 2008, p. 491). That is false. The ADA protects "qualified individuals with disabilities." For actual, present coverage by the ADA, then, one must have an actual, present disability. And so it should be. Obviously, everyone with a disability has an actual disability. One can no more be disabled by a nonactual disability than one can be clothed in nonactual clothing or bitten by a nonactual dog. In the ADA's terms, (C) covers people with the actual disability of being falsely believed to have an SLI. Likewise, everyone now disabled[57] has a present disability, including claimants under (B). They have a present history of a past SLI—a present disability, not merely a record of a past disability.

As an example of contradiction, a text in the *Nutshell* series says that the ADA

> define[s] an individual with a disability as one who (1) has a physical or mental impairment that substantially limits one or more of the individual's major life activities; (2) has a record of such impairment (*e.g.*, someone with a past history of cancer); or (3) is regarded as having such an impairment (*e.g.*, an individual who has been misclassified as disabled or who is treated as being disabled when in fact he or she is not disabled).[58]

This quotation makes no sense, since it says that someone can have a disability by virtue of being misclassified as disabled. Obviously, one cannot be disabled by being wrongly thought to be so; then the classification would not be wrong. At most, in the ADA's terms, one can be disabled under (C) by being wrongly thought disabled under (A).

To escape contradiction, one can, as always, make 'disability' ambiguous between two senses: one covering (A) alone, the other covering (A)–(C). But that is too confusing. Far better is to use two distinct terms. One can use 'disability' naturally and narrowly, as textbook writers do, to cover only actual present SLI's in clause (A). One must then rename both the statute and the protected class by some second term. In effect, textbooks treat the ADA as saying what it should have said, rather than what it did say. Besides its naturalness, the virtue of this approach—and one reason I stress these seemingly trivial mistakes[59]—is that it lays bare a premise the ADA's usage hides: that the three situations (A)–(C) have some unifying similarity for ethics or public policy. That is scarcely obvious. Alternatively, one can keep 'disability' broad, as in the ADA, and stick to a different term, like 'handicap' or 'substantially limiting impairment', for conditions in clause (A). I do not favor this second option, since it severs the link between disability and inability. A person disabled only under clause (C) has no intrinsic inability of any kind. His "disability" is only other people's false beliefs, which may not handicap his activities if he does not need their help. Of course, this feature will please activists who say that no disability is intrinsic to the individual. But (B) violates the rule in any case, since a past SLI need not be a present inability of any kind—not even a social handicap, if no one knows or cares about your history.

Clearly, then, regardless of terminology, as long as federal discrimination law includes categories (B) and (C), it protects some people who have no medical disorder. But what of 'impairment' in clause (A)? Can we say that, in the ADA, (i) an impairment must be a pathological condition? And is the converse also true, that (ii) every pathological condition is an impairment, so that impairment = pathological condition? I shall argue that the equation is nearly correct.

From EEOC regulations and court decisions, it seems that either (i) is true, if some errors in the regulations are corrected, or, at worst, 'impairment' is slightly broader than 'pathological condition'. The EEOC regulations define "physical or mental impairment" as:

(1) any physiological disorder, or condition, cosmetic disfigurement, or anatomical loss affecting one or more of the following body systems: neurological, musculoskeletal, special sense organs, respiratory (including speech organs), cardiovascular, reproductive, digestive, genito-urinary, hemic and lymphatic, skin, and endocrine; or
(2) any mental or psychological disorder, such as mental retardation, organic brain syndrome, emotional or mental illness, and specific learning disabilities.[60]

Here part (1) is obviously wrong, since it includes all normal conditions. Having two legs, two eyes, a beating heart, and weight above one ounce are all "conditions" that affect one or more of the systems listed, but impair nothing. To fix this problem, the EEOC then states that conditions are not impairments if they are in the "normal range" and do not result from a physiological disorder [29 C.F.R. pt 1630, app.

Sec. 1630.2(h)]. This is either inadequate or redundant. If 'normal' means "statistically normal," then unusually great strength or endurance or intelligence would still be an impairment. On the other hand, if 'normal' means "medically normal," the first clause is the same as the second, assuming 'pathological condition' and 'disorder' coincide. What the EEOC writers should have done is simply to insert the adjective 'pathological' before 'condition'. Also unclear is why the text lists "body systems." Is there some human organ system dysfunction of which is not an impairment? If not—if the list is intended to include all normal physiological systems—then what these are is a scientific issue on which federal rules are neither necessary nor possible.

A reasonable conjecture is that the agency rules aimed at least to make all pathological conditions impairments, verifying (ii). Whether anything else is an impairment for U.S. law may depend on whether purely cosmetic defects, or purely structural disorders in general, and HIV infection are pathological. Revised as I suggest, the agency rules still seem to cover "disfigurement" or "anatomical loss" without dysfunction. (I have found no interpretative guidance on these terms.) Clearly, one can be remarkably ugly—think of Charles Laughton—without having any medical disorder.[61] If mere ugliness can be "disfigurement," then impairment goes beyond pathological conditions. As for anatomical loss, I concede that medical reference works often include purely structural disorders.

An interesting case is HIV infection, the topic of a Supreme Court case. In *Bragdon v. Abbott* [524 U.S. 624 (1998)], a five-person majority held that HIV infection is always an impairment, even in its initial and its later "asymptomatic" phases.[62] But the majority's reason is precisely that HIV infection at any stage is a medical disorder. It assumes, in the initial stage, an "immediate" "assault on the immune system" and "a sudden and serious decline in the number of white blood cells." This seems to imply immunologic dysfunction. The Court says the term "asymptomatic phase" is "a misnomer," since "clinical features persist throughout, including lymphadenopathy, dermatological disorders, oral lesions, and bacterial infections." All of these are pathological conditions that can be coded as such in ICD-10. "In light of these facts," the court concludes, "HIV infection must be regarded as a physiological disorder with a constant and detrimental effect on the infected person's hemic and lymphatic systems from the moment of infection." Thus, it seems that either the Court majority rejected ICD-10's view of the medical status of some HIV infections, or it believed that, as a matter of fact, in every case of the Z-category "asymptomatic HIV infection status" (Z21), some disorder in B20-B24 is actually present as well.

Under both agency and court interpretations of the ADA, then, it may be that all impairments are either pathological conditions—if we follow the usage of medical sources that admit purely structural pathology—or, at worst, nonpathological "disfigurements." To the converse, (ii)—that every pathological condition is an impairment—there remain two obstacles, both removable, in court rulings and the ADA's text. The less serious objection comes from (B). Virtually everyone has had at least one bout of infectious disease which substantially limited nearly all major life activities. Many childhood diseases like measles, mumps, and chicken-pox and adult diseases like flu leave one bedridden for up to a week. If such a history does not suffice for (B), that can only be because a week of disease is too short to qualify. But if a minimum time is part of the meaning of 'impairment,' then 'impairment' does not mean "medical disorder," since these short-term, self-limiting illnesses are genuine disease. This obstacle disappears if minimum duration limits not 'impairment', but 'substantial'. To be a disability under (A), an impairment might have to last a substantial time, regardless of its momentary severity. That is what the Supreme Court said in *Toyota*: for an impairment of manual ability to be substantially limiting, its effect must be "permanent or long-term."[63]

A second obstacle to (ii) is that the ADA explicitly excludes from disability various well-recognized medical disorders. That is apparently because, though they are long-lasting ones which seem substantially to limit major life activities, Congress did not wish their bearers to enjoy ADA protection. One example is addiction to illegal drugs. Although addiction to legal drugs, such as alcohol, can apparently be a disability, addiction to illegal ones like marijuana, cocaine, or heroin cannot.[64] Yet the legal status of a drug is irrelevant to whether addiction to it is a medical disorder.[65] Moreover, if anything, addiction to an illegal drug more severely limits major life activities than addiction to a legal one, because of the drug's extra cost, the unreliability and inconvenience of its supply, and the threat of prison. Also denied by the ADA to be disabilities are "sexual behavior disorders," including pedophilia, exhibitionism, and voyeurism; and pyromania, kleptomania, and compulsive gambling.[66] All of these are psychiatric disorders in *DSM*. Can one sensibly deny that they ever substantially limit a major life activity? If reproduction can be a major life activity, as the Court said in *Bragdon*, it is hard to see why sexual life cannot. Indeed, intimate relationships enjoy constitutional significance under the line of privacy cases from *Griswold* and *Roe* to *Lawrence*.[67] Yet a man's pedophilia is a heavy burden on his adult sexual life, just as his homosexuality is a heavy burden on his heterosexual relationships. Similarly, if work can be a major life activity, it is hard to see why keeping oneself solvent cannot; yet compulsive gambling, if it exists, is a major burden on one's solvency. One could, of course, argue that a gambling compulsion can be resisted by force of will. But that seems equally true of alcohol addiction, and many purely physical limitations, such as pain, weakness, or fatigue, can also be largely overcome by willpower. Anyway, even if one can resist acting on abnormal sexual desires, one cannot replace them at will with normal ones, and the absence of the latter impedes any normal sexual relationship.

For the ADA's laundry list of excluded disorders to fit part (A) of its definition of disability, it must be that either (i) these disorders are not impairments, or (ii) they never substantially limit a major life activity. Since I see little appeal in (ii), one reading of the ADA is that the excluded disorders are not impairments. But this interpretation poorly fits the text in one way: in excluding its specific conditions, the ADA explicitly says that homosexuality and bisexuality are not disabilities because they are not impairments (12211(a)), whereas it only says the disorders I listed are not disabilities (12211(b)). This different treatment, given that DSM-III dropped homosexuality *per se* from the list of psychiatric disorders, suggests a desire by ADA writers to identify impairments with disorders, as courts and agencies also seem to have done. What is clear is that Congress did not wish to explain why the excluded disorders are not disabling. It is not clear how to reconcile the section-12211 exclusions with the general definition of disability in 12102.

Probably the best solution is to assume that the ADA's later sections restrict its general definition: in other words, to add to 12102(2) the implicit proviso "except as hereinafter provided by sections 12210 and 12211." The Rehabilitation Act has exactly such a clause at the parallel place, so the absence of one in the ADA may be an oversight.[68] If so, one can claim that all medical disorders are ADA impairments. Such a scheme, though logically satisfactory, is not morally so, since it still leaves obscure what the disorders excluded from disability have in common, and so what the remaining ones might share. It would be better to revise the statute so as either to include all the excluded disorders, or to cite in (A) a general property excluding them. Either course would clarify the ADA's moral basis, if any.[69] At any rate, it seems best to conclude that, under current US (though not UK) law, all disorders are impairments, and all impairments are either disorders (including purely structural pathology) or, perhaps, disfigurements.

It is likely, finally, that at least one component of the ADA's concept of disability—major life activity—is not purely medical. Ultimately, the sciences of biology and anthropology can fix some list of typical human activities. But no plausible such list will include all the things American courts have called major life activities. Playing intercollegiate sports[70] is neither a basic human biological capacity nor an activity typical of human societies. Even paid work is an economic, not a biological, category; and many societies either have no such institution, or exempt or block large classes of adults, especially women, from doing it. There is no reason to think that American courts which count working as a major life activity[71] would change their mind, for female plaintiffs, on evidence that most human adult females do not do paid work. So "major life activity," in ADA case law, does not seem to be a scientific category, medical or otherwise. Moreover, if the *Bragdon* minority is right that reproduction is a major life activity for some people but not others, depending on how they view it, then its status as such rests neither on scientific nor on medical judgment. In my view, an individual's medical status cannot vary with his own evaluation of his physical condition.[72] So, if I am right, judges often interpret the ADA's definition of disability to include nonbiomedical elements. Finally, even if what is a major life activity for human beings were a biomedical fact, it is hard to see how what limitations of them are substantial could be.

4.4 Two Approaches to Disability Advocacy

I end by briefly contrasting two approaches to disability advocacy, one consistent with my own philosophy of medicine, the other not. The inconsistent one is Ron Amundson's strategy (2000): to deny the scientific objectivity of pathology judgments. Comparing normality to race, he suggests that disease judgments, at least of nonlethal disorders, are mere bigotry—prejudice against alternative, "unfashionable" modes of function.

> Like the concept of race, the concept of normality is a biological error.... [T]he doctrine of biological normality is itself one aspect of a social prejudice against certain functional modes or styles. The disadvantages experienced by people who are assessed as 'abnormal' derive not from biology, but from implicit social judgments about the acceptability of certain kinds of biological variation.[73]

Amundson's examples, seemingly intended to illustrate his thesis, include blind and paraplegic human beings and a bipedal goat. So Amundson apparently means to deny that vision is a normal function of the human eye, that walking is a normal function of human legs, and that normal goats are quadrupedal. Although the main explicit target of his essay is my work, his views also contradict most recent philosophical writing on biological function.[74] Given the race analogy, they discredit not only the theory, but also, one would suppose, the practice of medicine.

By contrast, the usual approach to disability advocacy rests not on science, but on ethics and social philosophy. This effort, since it grants the scientific reality of disease, in no way contradicts my work in philosophy of medicine. In fact, as several writers note, it can draw support from some elements of my position.[75] For example, I stress the gap between theoretical and practical judgments, and the limited, purely instrumental, value of health. First, disease judgments entail nothing about treatment; *a fortiori*, they do not settle how to apportion our efforts between treating diseased individuals and changing their environment. Second, there is no intrinsic value in normality. Although its instrumental value enjoys a strong presumption, pathology is occasionally preferable to health, and, for almost any disease, one can imagine a special environment where it is advantageous.[76] Nothing in my analysis of health, then, is likely to block campaigns to modify human environments so as to make assorted pathology less burdensome.

Actually, as far as I can see, my views on philosophy of medicine fit even the most radical position on disability, the social model. In place of Amundson's denial of basic functional facts about mammalian physiology, the usual social-model theorist is, again, best seen as making an ethical, not a scientific, claim. A social model consistent with my health analysis would assert, not that normal human functional ability does not exist, but that ethics forces us to redesign the human environment, at any cost, to make it irrelevant. I am not persuaded that social-model theorists have yet said anything either true or useful. Still, it is important to distinguish Amundson's version of it from the usual type. Even disability radicalism need not rest on weird science. It can rest on weird ethics instead, which may, in time, prove more persuasive.

Notes

1. Barnes, Mercer, and Shakespeare (1999, p. 6) cite a "growing consensus on the oppressive implications of the term 'handicap', mainly because of its historical allusions to 'cap

in hand', begging and charity...." That might be a plausible etymology for the word 'capi-hand', but there is no such word. Old posters urged employers to "Hire the Handicapped," not to "Choose the Capihanded." Not surprisingly, the *Oxford English Dictionary*, from its first edition (Murray, 1901, p. 62) to its current online version—like every dictionary I have seen which gives any etymologies—derives 'handicap' from "hand in cap," referring to an antique kind of wager where players concealed their hands. The term spread to horse racing, where, to this day, a handicap is extra weight placed on faster horses to equalize slower ones' chances. Far from evoking beggary, the term's history suggests that a handicap is something an unusually strong individual can win by overcoming. So Barnes et al. offer a fine example of taking offense at imaginary slights.

2. *Online Oxford English Dictionary*, 2008, entry "disability." The quoted definition is part a of a first sense, which continued: "b. An instance of this. ... c. Pecuniary inability or want of means."

3. Ibid. "Incapacity in the eye of the law, or created by the law; a restriction framed to prevent any person or class of persons from sharing in duties or privileges which would otherwise be open to them; legal disqualification."

   Recently, the online OED added a third sense: "a physical or mental condition that limits a person's movements, senses, or activities." This still seems overbroad in the same ways as sense 1 (inability), as well as for other reasons that will appear.

4. This is specifically denied to be a disability in the first edition of the WHO classification (1980, p. 1547).

5. My discussion of these conceptual frameworks is heavily indebted to Jette and Badley (2002). A briefer survey of disability models is Rondinelli and Duncan (2000). Especially valuable was Nagi (1991).

6. Nagi (1991, p. 317). A similar claim is made by ICIDH (1980, p. 31).

7. ICIDH's actual list of disabilities, however, includes some limitations of social-role activities, such as work and family life. Examples are "tolerance of work stress" (D76) and virtually all of "family role disability" (D17). Consequently, Jette and Badley (2002, p. 193), echoing Nagi's own criticism (1991, p. 325), conclude: "The ICIDH-1 term *disability* ... bridges the Nagi concepts of functional limitation and disability."

   Actually, to some degree, ICIDH's conceptual introduction bases the disability/handicap distinction not on personal vs. social, but on fact vs. evaluation (1980, pp. 28–29). This is a different distinction; e.g., evaluation can be by the impaired person, as ICIDH itself notes. As with 'impairment', below, the ICIDH view of 'handicap' straddles at least two quite different analyses.

   Edwards (1997, p. 594) and Nordenfelt (1997, p. 609) find two or three more contrasts within ICIDH's disability/handicap distinction. But that is partly because each of them mis-states ICIDH's text. Edwards (1997, pp. 590–591) misreads the ICIDH "survival roles" (1980, pp. 38–39) as aspects of disability, rather than of the disadvantage of handicap. This error ruins three paragraphs of his criticism (Edwards, 1997, pp. 596–597). Nordenfelt (1997, p. 609) says that disabilities are "simple," handicaps "complex," but ICIDH says that disability involves "compound" (1980, p. 28) or "composite" (1980, p. 143) activities. Nordenfelt's example of inability to lift one's arm may be a "paradigm example" (ibid.) of his own idea of disability, but in ICIDH it seems to be an impairment (1980, p. 98).

8. Most of his criticisms are summarized in footnotes 18 and 29, *infra*.

9. Barnes and Mercer (2003, p. 15). More exactly, what correspond to 'disability' and 'handicap' in ICIDH are "activity limitation" and "participation restriction" in ICF (2001, p. 8). However, most dimensions of ICIDH's contrast are now gone; see below.

10. It is unclear to me how ICF is an improvement on ICIDH except as to impairment, as noted below. As a conceptual scheme (not a classification), ICIDH looks superior to ICF, and Nagi is the best of all.

    What is clear is that the WHO writers are oversensitive to criticism. Regarding terminol-ogy, they say that "WHO confirms the important principle that people have the right to be called what they choose" (2001, p. 242). Most people accept this rule for proper names, like 'Muhammad Ali' or 'World B. Free'. Some also extend it to quasi-proper names of classes, like 'Native American' or 'African-American'. But there is no reason to think classes of people have a right to choose their own descriptive predicate, such as 'person with disability'. E.g., philosophy professors have no right to be called "persons of profundity," "moral paragons," "natural rulers," etc., no matter how much they might enjoy these descriptions. And, of course, before dropping a term ('handicap') marking a basic distinction, because of a false etymology absurd on its face, one might have hoped that someone at WHO would look in a dictionary.

11. For Nordenfelt's defense of this distinction—relying on the difference between basic and generated acts—see his (1997), replying to Edwards (1997).

    ICIDH's disability/handicap distinction is not the only one in clinical textbooks. An AMA textbook uses 'handicap' for compensated impairments, such as having to wear an artificial leg or sit in a special chair—or, presumably, needing eyeglasses (Demeter & Andersson, 2003, p. 5). In such cases, the text says, there is handicap without disability. This seems to be either the reverse terminology from ICIDH's or a wholly different usage.

12. I use the neutral term "associated with" to bracket causal controversies. Advocates of the social model claim that impairment does not cause disability, but society does. It is not clear on what analysis of causation these claims can both be true; for some discussion, see Cox-White and Boxall (2009). In this essay, I ignore all metaphysical issues about disability, including causation, and all empirical causal questions as well.

13. In this category are Taber (1940) and *Stedman's Medical Dictionary* (1953). On the other hand, neither of these works lists 'disorder' either, which has at least come to be a central foundational term in medicine today.

14. E.g., *Taber's* 20th ed. (2005); *Stedman's* 5th ed. (2005).

15. ICF (2001, p. 13). This statement is essentially unchanged from the first edition, ICIDH (1980, p. 27).

16. See Boorse (1977, pp. 550–551; 1987, pp. 362–364; and 1997, pp. 41–42). Without a broad disease concept, the medical cliché that health is the absence of disease must fail.

17. Perhaps it is because, in WHO's corresponding classification of medical disorders (ICD-10), a missing leg would fall under section Q (if congenital) or section S or T (if externally caused), sections which, unlike A-N, do not refer to "diseases" in their titles or "disorders" in their texts. This is a bad argument, however, for, in medical usage, 'disorder' undoubtedly includes Q-coded genetic syndromes such as Marfan's (Q87.4), Down's (Q90), or Turner's (Q96).

    To be fair to disability writers, I should note that many philosophers of medicine, too, fail to see that medicine has a general theoretical term—*pathological condition*—covering diseases, injuries, static defects, poisoning, environmental effects, etc. A notable exception is Reznek (1987, p. 65 ff.). Wakefield may be using *disorder* for the same purpose; see, e.g., his (1992, 1999).

18. ICIDH (1980, p. 26). In turn, a pathological state is "objectified" into *disability* when it changes the person's "functional performance and activity," and "socialized" into *handicap* when it leads to a social disadvantage (ibid.).

    Nagi (1991, pp. 321, 324) criticizes ICIDH's use of all three of these terms. As to "exteriorization," he finds it illogical to make the existence of an impairment depend on anyone's awareness of it. As to "objectification," he says that impairment, functional limitation, and disability, like anything else, "can all be considered from objective or subjective viewpoints." His general criticism of "socialization," though unclear to me, concerns ICIDH's remarks on evaluation.

19. Indeed, ICF (2001, p. 12) explicitly says that "[i]mpairments have been conceptualized in congruence with biological knowledge at the level of tissues or cells and at the subcellular or molecular level," and that its classification may later be extended to those levels. Possibly, of course, ICIDH merely meant to claim that impairments were at organ level *or below*— i.e., not personal or social.

20. Nagi used "abnormality or loss" (1991, p. 322); ICIDH "loss or abnormality" (1980, p. 47).
21. Namely, a pathological state is one of statistically species-subnormal biological part-function (Boorse, 1997, p. 1), relative to a person's sex and age. One must, of course, take care to distinguish the pathological (dysfunction) from the pathogenic (a cause of dysfunction) and the pathodictic (a sign of dysfunction).
22. Examples in ICIDH may include abnormal trunk hair (83.4) and accessory nipples (67.2).
23. Boorse (1977, pp. 565–566; 1997, p. 44). Actually, it is far from clear how to incorporate structural abnormality into a definition of disorder. Some statistically atypical structure, such as a very large cerebrum, is not pathological. Most writers agree that, properly analyzed, 'function' has an inherent directionality that lets us call all statistically subnormal function pathological. But at either statistical extreme, some structural abnormality is pathological (missing arms; extra arms) and some is not (scanty leg hair, large cerebrum).
24. ICF, p. 238. Annex 4 (1980, pp. 238–241) is a revised and expanded version of ICIDH, p. 31, which had many of the same examples.
25. The Z-category covers "circumstances other than a disease, injury or external cause classifiable to categories A00-Y89" (ICD-10, p. 1125). "Asymptomatic HIV infection status" (mere HIV-positivity) is Z21, while the categories for "HIV disease" are B20-24.
26. This conclusion confirms David Wasserman's assumption (2001, p. 223) that my own essays offer an analysis of 'impairment'. But, because of my next point, Wasserman's decision in this essay to call his topic "people with impairments," rather than "people with disabilities" (p. 220), departs from most disability literature. The class of people with impairments is just the class of people—or very nearly so.
27. 42 U.S.C. §12101(a)(1), (7). The quoted phrase, from *U.S. v. Carolene Products*, 304 U.S. 144, 152 n. 4 (1938), is the Supreme Court's classic description of what kind of group can suffer "prejudice" that, in later constitutional discrimination law, triggers "strict scrutiny" of their legislative classification. The Court, however, does not count disability as a "suspect classification" (like race), requiring strict scrutiny, or even as a "quasi-suspect classification" (like sex), requiring intermediate scrutiny. *City of Cleburne v. Cleburne Living Center, Inc.,*105 S.Ct. 3249 (1985).
    Justice Stevens (joined by Breyer), dissenting in *Sutton v. United Air Lines, Inc.*, 527 U.S. 471, 495 (1999), jokes that Congress's own "myopia" may have kept it from realizing that "its definition of 'disability' might theoretically encompass ... perhaps two or three times" the 43 million—namely, everyone who wears glasses. Stevens' reason is that he believes Congress assumed—as its key committees explicitly stated—that, as with artificial legs, the limiting effects of impairments must be judged in their untreated or uncompensated state. Stevens' view suggests that most Americans, including nearly everyone past late middle age, may be disabled. However, a unanimous Court, three years later, again cited the 43 million figure to argue that the ADA's terms "need to be interpreted strictly to create a demanding standard for qualifying as disabled." *Toyota Motor Manufacturing, Kentucky, Inc. v. Williams*, 534 U.S. 184,197 (2002). In the ADA Amendments Act of 2008, passed after this paper was written, Congress endorsed Stevens' view of compensated impairments, as well as broadening ADA coverage in several other ways.
    Of course, by the ADA in any form, certainly not all impairments are disabilities, since, to be so, they must also "substantially limit" a "major life activity." See Section 4.3.2.
28. Whitfield (1997), cited in Doyle (2008, p. 15, n. 18).
29. Perhaps Nagi's most telling criticism of ICIDH is that it has no clear concept of "performance at the level of the organism" (1991, p. 325). That is because, first, as we saw, its examples of disability are "a mix of social and organismic performance." At the same time, it counts functional limitation as "an aspect of impairment" (WHO, 1980, p. 28), which is itself "conceptualized at the organ level" (Nagi, 1991, p. 325). Perhaps, however, ICIDH and Nagi mean something different by functional limitation, since it is at the organ level for ICIDH, at the organism level for Nagi. As attentive readers will notice, not one of the four key nonmedical terms in disability literature—'impairment', 'functional limitation', 'disability', and 'handicap'— has a consistent usage among leading authorities.

30. In analyzing life and death, some philosophers try to distinguish functioning of an organism as a whole from merely local organ function. It is interesting how hard this distinction is to draw. For example, Culver and Gert end up holding that respiration is, but circulation is not, a function of an organism as a whole (1982, pp. 186, 188, 190–191). This result, though needed for their concept of brain death, seems implausible.

31. Some combination of these ideas is called "functional capacity" in textbooks. E.g., the AMA textbook uses the term 'functional capacity assessment' for "the spectrum of tests [of] an organ or organ system in performance of its basic function" (Demeter & Andersson, 2003, p. 689). See also Rondinelli and Katz (2000, chap. 5).

    One might think a focus on activities the best way to define disability, if one thinks of ability as ability to act. But many processes often called abilities, such as sensation, perception, speech comprehension, and various cognitive skills, are no more voluntary acts than serum-glucose regulation or immune defense. And serious performance deficits in these abilities are classic examples of disabilities. To define disability via "basic acts," as Nordenfelt (1987) suggests, may also be to make it a nonbiological concept. After all, many organisms, from lower animals to all plants, exhibit nothing like voluntary action.

32. On causal-role analyses like Cummins's (1975), an organ's function is its disposition causally to contribute to the outputs that interest us of a complex containing system. In most biological contexts, the main outputs of interest are the organism's survival and reproduction. These are also organism goals, according to my own analysis (1976), to which an organ's functions are causal contributions. For etiological analysts, from Wright (1973) to more recent writers such as Neander (1991) and Millikan (1984), the functions of a trait are those effects by which natural selection shaped it, and so are again past contributions to its ancestors' reproduction. Among major types of analysis, only value-centered ones, such as Bedau's (1992), award a trait a function otherwise than by its external effects. (For example, if conscious experience is good for an organism, the brain might have purely internal functions in supporting this good.) For recent summaries of the function debate, see Boorse (2002, pp. 63–68), for a few pages, and Wouters (2005) for a whole paper. Book-length surveys are Melander (1997) and Nissen (1997). Three recent anthologies are Allen, Bekoff, and Lauder (1998), Buller (1999), and Ariew, Cummins, and Perlman (2002).

    It is worth mentioning that common clinical references, especially in mental-health professions, to people's "daily functioning" risk some confusion. Parts and processes within organisms have biological functions, as do their species-typical activities. But whole organisms, including human beings, have no biological functions, except in an ecological usage irrelevant to health concepts. For some discussion, see Boorse (1976, pp. 84–85).

33. *AMA Guides* (2001, p. 10). It is, of course, a formula for the probability that at least one of two independent events occurs.

34. Despite its introduction's references to "pathological processes" (1980, p. 27) and "disease or disorder" (1980, p. 30), ICIDH codes some normal conditions as impairment (pregnancy, 99.0, p. 116) or disability (left-handedness, 65, p. 167). Both ICF and American and English discrimination law also admit normal disabilities (see below), while the *Guides* recognizes gross impairment without disability (2001, p. 8, Figure 1-1).

35. ICIDH lists a "handedness disability" of "being a sinistral in a predominantly dextral culture" (1980, p. 167, D65). Left-handedness is not on its list of impairments, nor a disorder in ICD-10. In ICF, of course, all activity or participation restrictions are disabilities, so the stigmatized former mental patient (2001, p. 239) or wrongly suspected HIV carrier (2001, p. 240) is disabled despite being currently healthy. Both also have a disability under the ADA's "record of" and "regarded as" clauses, respectively; see Section 4.3.2 below.

36. E.g., Clouser, Culver, and Gert (1997). However, these writers' definition of 'disability' is even more clearly stipulative than their definition of 'malady', which at least purports to capture an existing medical concept. See note 40, *infra*.

37. Frisancho (1993, pp. 72–73). Interestingly, by contrast, adult immigrants adapt to dry desert heat as fully as natives (Frisancho, 1993, p. 189).

38. Moran's Figure 6.7 (2000, p. 161) gives the two values for maximum VO_2 as approximately 52 and 36 ml/kg/min. He says the data "suggest that there might well be a critical age at which migration must occur if the individual is to achieve a high VO_2 as an adult" (2000, p. 160).

39. A psychological counterpart to these examples is language acquisition. Nearly any young child can learn to speak any human language like a native. But many transplanted adults cannot learn to speak a new language without accent. One can imagine such an accent's limiting someone's ability to work.

40. Some philosophers of medicine offer analyses of disease-like concepts that embrace pregnancy. An example is Clouser et al. (1997), who are willing to count pregnancy (pp. 205–207), menopause (p. 207), and teething (p. 208) as "maladies" while conceding their normality. I believe that such analyses either do not aim at the basic category of theoretical medicine—pathological condition—or are incorrect. See my remarks in the same volume (1997, pp. 43–44).

41. Mayerson (1962, p. 388) says that only those jobs count "for which the insured is reasonably fitted by education or training." Thus, "An engineer who has lost his legs but can still sell programs outside the football stadium would be considered to be totally disabled even under the 'any occupation' definition."

42. E.g., I leave out the Individuals with Disabilities Education Act, which applies only to minors. For some discussion, see Blanck, Hill, Siegal, and Waterstone (2004). Also, I have not investigated the rules by which wounded soldiers are disabled from combat duty.

43. This has long been the view of the AMA, whose Committee on Medical Rating of Physical Impairment said in 1958 that disability was "not a purely medical condition," and its evaluation "is an administrative, not medical, responsibility" (1958, p. 3).

44. Of course, someone might object that this pitcher can still do the job—just extremely badly, so badly that his team is guaranteed to lose every game he pitches, unless aided by blind or corrupt umpires. Such a view only strengthens the point of the paragraph.

45. On the ADA provision see Tucker and Milani (2004, p. 29). This example also illustrates the frequent use, in disability law, of degree-vague nonmedical terms like 'reasonable', 'substantial', etc. As another example, a law dictionary states: "total disability to follow insured's usual occupation arises where he is incapacitated from performing any substantial part of his ordinary duties, though still able to perform a few minor duties and be present at his place of business" (*Black's Law Dictionary* 1979, p. 415).

46. For purposes of "own-occupation" insurance coverage. But the pitcher is not disabled for the ADA; see 29 CFR XIV (7/1/06) §1630 App. at 369.

47. In other words, the objection is that *Heart of Atlanta Motel v. U.S.*, 379 U.S. 241 (1964), was wrongly decided. This case is inseparable from deep issues about constitutional structure, especially whether the federal government may use enumerated powers as pretexts to achieve other, unauthorized, goals. In general, the Court holds that it may, unless the law threatens fundamental rights grounded in amendments other than the 9th and 10th, or other explicit constitutional limits on federal power. For some discussion, see Epstein (1992, pp. 135–143).

48. Regarding the CRA, this claim rests on the belief that race and sex discrimination were already widely condemned in 1964, and would have continued to wither without the parts of the CRA touching private discrimination. That is the view of Charles Murray (1997, pp. 87–88). Epstein (1992) reaches similar conclusions. He also makes the economic case for bad effects of the CRA, even apart from its interpretation by a unanimous Court, in *Griggs v. Duke Power Co.*, 401 U.S. 424 (1971), to ban "disparate impact" instead of "disparate treatment," a result he calls "a travesty of statutory construction" (p. 192). As to disability-discrimination law, in particular, it has been argued that the ADA reduces employment of disabled people by making it look more expensive; see Jolls (2000).

49. For the ADA's description of disabled people as a minority, see text accompanying note 27. By contrast, the CRA, as written, clearly protects people of all races and both sexes, and it is widely agreed that congressional consensus on this feature was crucial to its passage. Only

later did Supreme Court majorities in *Bakke*, 438 U.S. 265 (1978), and *Weber*, 443 U.S. 193 (1979), willfully misinterpret the CRA to permit reverse discrimination.

50. Colker (2005, p. 18). See also her explanation (p. 97) of how the ADA's coverage is "radically different" from the 1964 CRA's. "The ADA is built on an 'antisubordination' notion for protected classes under which Congress sought to assist a historically disadvantaged class: individuals with disabilities." By contrast, the CRA rests on an "antidifferentiation" theory. For more on the contrast between these two approaches, see Bagenstos (2000).

51. *Sutton v. United Air Lines, Inc.*, 527 U.S. 471, 504 (1999) (Stevens, J., dissenting). Feldblum (2000) concludes that the ADA ought to prohibit discrimination on the basis of all impairments, and the ADA Amendments Act of 2008 looks like a significant step in this direction.

52. The contrast between the two approaches is well drawn by David Wasserman (2000). As he notes, there is no obvious reason why Stevens' involuntary characteristic must be an inability, let alone a medically abnormal one. Wasserman's own view, going beyond Feldblum, is that ADA coverage should extend even past impairments which are not substantially limiting.

> A revised statute ... should protect anyone with a disfavored physical or mental variation: it should apply to those who are overweight but not morbidly obese, short but not achondroplastic, unattractive but not disfigured, and "dull-witted" but not mentally retarded (ibid., p. 148).

Really, Wasserman's thesis is that the disability concept is irrelevant to discrimination ethics and law. This seems fatal to the hope that any antidiscrimination goal can define disability.

53. By the M'Naghten rule, insanity is mental disease which keeps the defendant ignorant of what he is doing or of its wrongness. By a control test, insanity can be mental disease that makes the defendant unable to control his conduct. The third popular rule, the ALI test, is essentially a disjunction of these older ideas. The fourth and broadest test requires only that the defendant's act be the "product of mental disease." But this product rule owes its influence to its use for eight years in only one American court, which later modified it in the direction of the ALI rule. There is little doubt that the bare product rule is unacceptable if literally applied. On tests of insanity, see LaFave (2003, pp. 368–401), Moore (1984), and Reznek (1997).

54. I ignore the difference between these two definitions in analyzing law, since it is hard to see how a clinically undetectable disorder could figure in a legal case. Even if a legal term meant 'disorder', one would need evidence, presumably clinical, to allege it in the legal process.

55. 42 U.S.C. §12102(2). This definition is repeated nearly verbatim from the Rehabilitation Act of 1973, 29 U.S.C. §705(20)(B) (2000).

As for the UK, its Disability Discrimination Act 1995 has a somewhat different version of (A). It defines a disability as "a physical or mental impairment which has a substantial and long-term adverse effect" on a person's "ability to carry out normal day-to-day activities" (s1 (1)). "Normal day-to-day activities" looks like a different concept from "major life activities." Moreover, by a supplementary schedule (Sch 1, 4(1)), disability requires an impairment in one of eight specific areas: mobility; manual dexterity; physical coordination; continence; ability to move everyday objects; speech, hearing, or eyesight; memory or ability to concentrate, learn or understand; perception of the risk of physical danger.

The UK law contains nothing like the ADA's (C). It includes a provision like (B)—namely, s2(1)—but it seems to avoid the confusions I discuss in the text, since it merely says that various parts of the law apply to a person with a past disability as if it were present.

56. Colker (2005, p. 101). The Supreme Court, per Justice O'Connor, has said almost exactly the same thing: "to fall within this definition one must have an actual disability (subsection (A)), have a record of a disability (subsection (B)), or be regarded as having one (subsection (C))" (*Sutton v. United Air Lines, Inc.*, 577 U.S. 471, 478 (1999)). This escapes falsity by pure logic: if to have p one must have q, then it is true that to have p one must have q, r, or

s. But the court is confused in labeling ADA's clause A "actual disability," B "record of a disability," and C "regarded as having a disability."

Nevertheless, these are extremely common section headings in books on US discrimination law, such as Blanck et al. (2004, p. 3–1) or Zimmer et al. (2008, p. 492). An honorable exception is Weber (2007, pp. 25, 36, 37).

57. No law or legal writer I have seen distinguishes between "having a disability" and "being disabled," so, in this section, I do not either.

58. Tucker and Milani (2004, p. 16). I am not being unfair to these authors, since they later write: "The third prong of the definition of an individual with a disability is intended to protect individuals who do not have, and may never have had, a mental or physical disability within the meaning of the law..." (p. 23).

59. On occasion, like any conceptual confusion, errors over the ADA's term 'disability' also defeat clear reasoning. An example is Justice Stevens' criticism, in *Sutton*, of the majority's claim that "[a] 'disability' exists only where an impairment 'substantially limits' a major life activity, not where it 'might,' 'could,', or 'would' be substantially limiting if mitigating measures were not taken" (527 U.S. 482). Stevens notes that the first clause of this sentence would eliminate (B) and (C) as kinds of disability. Although this point is correct, it leaves the majority's argument untouched. Since the only issue in the case was whether the Sutton sisters' myopia was an SLI, the argument needs only the premise that disability under (A), viz., SLI, is always present and actual—not disability in general.

60. 29 CFR §1630.2(h)(1) (1998), quoting HEW regulations interpreting the 1973 Rehabilitation Act, 45 CFR §84.3(j)(2)(i) (1997). It is disputed how much legal force EEOC interpretations of basic ADA terms have; see Zimmer et al. (2008, pp. 520–522).

61. ICIDH includes "marked ugliness (*e.g.*, gargoylism)" as an impairment (82.8). Interestingly, both *Stedman's Medical Dictionary* (1976, p. 570) and ICD-10 list gargoylism only as an effect of mucopolysaccharidoses such as Hunter's and Hurler's syndromes. Neither book suggests that merely looking like a gargoyle is pathological in itself.

62. Also, none of the four dissenting justices denied this thesis, which was undisputed by the parties. The issues in the dissents are (i) whether reproduction is always a "major life activity," (ii) whether HIV infection substantially limits reproduction, and (iii) whether office procedures on an HIV-positive patient pose a "direct threat" to the dentist.

63. 534 U.S. 184, 198 (2002). The term 'permanent' seems redundant, unless it protects impaired persons who have not long to live. Otherwise, every permanent impairment is long-term. Conversely, 'long-term' is essential: there is no prospect of requiring every SLI to be permanent, since then clause (B) for past SLI's would be unnecessary. Surprisingly, in a later decision, an appellate court held that nine months was not a long term. *Pollard v. High's of Baltimore, Inc.*, 281 F.3d 462 (4th Cir. 2002). It is unclear what principle can resolve the tension between clause (B) and the desire to keep disabled people a minority. It is also unclear how the Court's view in *Toyota* fits its statement in *School Board of Nassau County v. Arline*, 480 U.S. 273, 281 (1987), that any hospitalization substantially limits a major life activity.

64. 42 U.S.C.A. §12114, 12210 (a). Zimmer et al. say that "the plain language" of §12114 (a) "indicates that an alcoholic who is currently using alcohol may be disabled under the ADA" (2008, p. 582).

In the UK, the corresponding restriction is to medically prescribed drugs, such as painkillers. Under 1996 regulations, addiction to legal recreational drugs does not qualify as an impairment (Doyle 2008, pp. 17–18). The objection still applies. Addiction to a drug prescribed by a doctor (or several doctors) can certainly be a medical disorder—e.g., addiction to a sedative (304.10) or phencyclidine (304.60) in DSM-IV-TR.

65. E.g., for alcohol and cocaine, DSM-IV-TR lists the same basic disorders: dependence (303.90; 304.20), abuse (305.00; 305.60), intoxication (303.00; 292.89), intoxication delirium (291.0; 292.81), and withdrawal (292.81; 292.0). Marijuana affords all but the last category. For a list of which drugs cause which types of disorders, see Table 1 (p. 193).

66. ADA, U.S.C. §12211(b). UK regulations, too, exclude pyromania, kleptomania, exhibitionism, and voyeurism even from impairment, let alone disability (Doyle, 2008, p. 18).
67. *Griswold v. Connecticut*, 381 U.S. 479 (1965); *Roe v. Wade*, 410 U.S. 113 (1973); *Lawrence et al. v. Texas*, 539 U.S. 558 (2003). The Ninth Circuit ruled that sexual relations are a major life activity in *McAlindin v. County of San Diego*, 192 F.3d 1226, 1234 (9th Cir. 1999).
68. Rehabilitation Act, 29 U.S.C. §705(20)(B): "Subject to subparagraphs (C), (D), (E), and (F), the term 'individual with a disability' means ...." Note that without such a clause, the ADA's provision in §12210 covering recovered illegal-drug addicts, but denying coverage to current ones, also contradicts the disability definition. Under that definition, being a past drug addict (B) or being regarded as a drug addict (C) cannot be a disability unless addiction itself is (A).
69. Interestingly, insanity law also makes dubious, possibly unprincipled, exclusions of specific conditions, especially psychopathy, from the insanity defense. For some discussion, see Reznek (1997).
70. This was held a major life activity in *Sandison v. Michigan High School Athletic Association, Inc.* (E.D. Mich. 1994) and in *Pahulu v. University of Kansas* (D.Kan. 1995).
71. Many courts have held or assumed that it is. For a general discussion of disputes about what is a major life activity, see Edmonds (2002); on work, specifically, see Rahdert (2000).
72. Except, of course, insofar as such evaluation is part of psychopathology, such as depression, delusion, etc.
73. Amundson (2000, pp. 33–34). The term 'unfashionable' is from pp. 48–50.
74. Most function writers, in distinguishing functions from mere effects, take biology to attribute to the traits of organisms normal functions—or, in an alternate, perhaps regrettable, usage, "proper" functions. This is true of goal-contribution analysts such as Nagel (1961), Boorse (1976), and Adams (1979), selected-effect theorists like Millikan (1984) and Neander (1991); and even many versions of the causal-role approach. Amundson and Lauder (1994) are among the few not to assume a notion of species-normal function. For general references on the function debate, see note 32, *supra*.
75. David Wasserman makes this point (2001, pp. 223–224). I also benefit here from unpublished work by Tim Lewens.
76. On the independence of abnormality and treatment judgments, see, e.g., Boorse (1975, p. 68; 1977, pp. 545–546). On the presumptive value of health, see Boorse (1975, p. 60; 1997, pp. 98–99). On desirable pathology, see Boorse (1975, p. 53; 1987, p. 369; 1997, p. 88). Regarding the merely instrumental value of health, I wrote:

> But there is presumably no intrinsic value in having the functional organization typical of a species if the same goals can be better achieved by other means. A sixth sense, for example, would increase our goal-efficiency without increasing our health; so might the amputation of our legs at the knee and their replacement by a nuclear-powered air-cushion vehicle (1975, p. 61).

The latter example is timely in a year that saw the International Association of Athletics Federations initially disqualify amputee runner Oscar Pistorius, whose prosthetic feet scientists found to be more efficient for running than normal ones.

References

Adams, F. R. (1979). A goal-state theory of function attributions. *Canadian Journal of Philosophy, 9,* 493–518.

AMA Committee on Medical Rating of Physical Impairment. (1958). Guides to the evaluation of permanent impairment. *Journal of the American Medical Association, 9,* p. 3.

Allen, C., Bekoff, M., & Lauder, G. (Eds.). (1998). *Nature's purposes*. Cambridge, MA: MIT Press.

APA (2000). *Diagnostic and statistical manual of mental disorders.* Washington, DC: American Psychiatric Association. *(DSM-IV-TR).*

Amundson, R. (2000). Against normal function. *Studies in History and Philosophy of Biological and Biomedical Sciences, 31,* 33–53.

Amundson, R., & Lauder, G. V. (1994). Function without purpose: The uses of causal role function in evolutionary biology. *Biology and Philosophy, 9,* 443–469.

Anderson, D. M. (2002). *Mosby's medical dictionary* (6th ed.). St. Louis, MO: Mosby.

Ariew, A., Cummins, R., & Perlman, M. (Eds.). (2002). *Functions.* New York: Oxford.

Bagenstos, S. R. (2000). Subordination, stigma, and "disability". *Virginia Law Review, 86,* 397–534.

Barnes, C., & Mercer, G. (2003). *Disability.* Cambridge, UK: Polity Press.

Barnes, C., Mercer, G., & Shakespeare, T. (1999). *Exploring disability: A sociological introduction.* Cambridge, UK: Polity Press.

Bedau, M. (1992). Where's the good in teleology? *Philosophy and Phenomenological Research, 52,* 781–806.

Black, H. C. (1979). *Black's law dictionary* (5th ed.). St. Paul, MN: West Publishing Co.

Blanck, P., Hill, E., Siegal, C. D., & Waterstone, M. (2004). *Disability, civil rights law and policy.* St. Louis, MO: Thomson (West).

Boorse, C. (1975). On the distinction between disease and illness. *Philosophy and Public Affairs, 5,* 49–68.

Boorse, C. (1976). Wright on functions. *Philosophical Review, 85,* 70–86.

Boorse, C. (1977). Health as a theoretical concept. *Philosophy of Science, 44,* 542–573.

Boorse, C. (1987). Concepts of health. In D. VanDeVeer & T. Regan (Eds.), *Health care ethics* (pp. 359–393). Philadelphia, PA: Temple University Press.

Boorse, C. (1997). A rebuttal on health. In J. M. Humber & R. F. Almeder (Eds.), *What is disease?* (pp. 1–134). Totowa, NJ: Humana.

Boorse, C. (2002). A rebuttal on functions. In A. Ariew, R. Cummins, & M. Perlman (Eds.), *Functions* (pp. 63–112). New York: Oxford.

Buller, D. J. (Ed.) (1999). *Function, selection, and design.* Albany, NY: SUNY Press.

Clouser, K. D., Culver, C. M., & Gert, B. (1997). Malady. In J. M. Humber & R. F. Almeder, (Eds.), *What is disease?* (pp. 175–217). Totowa, NJ: Humana Press.

Cocchiarella, L., & Andersson, G. B. J. (Eds.). (2001). *Guides to the evaluation of permanent impairment* (5th ed.). USA: AMA Press. *[AMA Guides].*

Colker, R. (2005). *The disability pendulum: The first decade of the Americans with Disabilities Act.* New York: NYU Press.

Cox-White, B., & Boxall, S. F. (2009). Redefining disability: maleficent, unjust, and inconsistent. *Journal of Medicine and Philosophy, 33,* 558–576.

Culver, C. M., & Gert, B. (1982). *Philosophy in medicine.* New York: Oxford.

Cummins, R. (1975). Functional analysis. *Journal of Philosophy, 72,* 741–765.

Demeter, S. L., & Andersson, G. B. J. (2003). *Disability evaluation* (2nd ed.). St. Louis, MO: Mosby.

Doyle, B. (2008). *Disability discrimination: Law and practice* (6th ed.). Bristol, UK: Jordans.

DPI (1982). *Proceedings of the First World Congress.* Singapore: Disabled People's International.

Edmonds, C. D. (2002). Snakes and ladders: Expanding the definition of "major life activity" in the Americans with Disabilities Act. *Texas Tech Law Review, 33,* 321–376.

Edwards, S. D. (1997). Dismantling the disability/handicap distinction. *Journal of Medicine and Philosophy, 22,* 589–606.

Epstein, R. (1992). *Forbidden grounds.* Cambridge, MA: Harvard.

Feldblum, C. (2000). Definition of disability under federal anti-discrimination law: What happened? Why? And what can we do about it? *Berkeley Journal of Employment and Labor Law, 21,* 91–165.

Frisancho, A. R. (1993). *Human adaptation and accommodation.* Ann Arbor, MI: University of Michigan Press.

Jette, A. M., & Badley, E. (2002). Conceptual issues in the measurement of work disability. In G. S. Wunderlich, D. P. Rice, & N. L. Amado (Eds.), *The Dynamics of Disability* (pp. 183–210). Washington, DC: National Academies Press.

Jolls, C. (2000). Accommodation mandates. *Stanford Law Review, 53*, 223–306.

LaFave, W. R. (2003). *Criminal law* (4th ed.). St. Paul, MN: Thomson (West).

Lange, J. L. (Ed.). (2006). *The Gale encyclopedia of medicine* (3rd ed.). New York: Thomson Gale.

Mayerson, A. L. (1962). *Introduction to insurance.* New York: Macmillan.

Melander, P. (1997). *Analyzing functions: An essay on a fundamental notion in biology.* Stockholm: Almqvist & Wiksell.

Millikan, R. (1984). *Language, thought, and other biological categories.* Cambridge, MA: MIT Press.

Moore, M. S. (1984). *Law and psychiatry: Rethinking the relationship.* New York: Cambridge.

Moran, E. F. (2000). *Human adaptability.* Boulder, CO: Westview Press.

Murphy, E. A. (1976). *The logic of medicine.* Baltimore, MD: Johns Hopkins.

Murray, J. A. H. (Ed.) (1901). *A new English dictionary on historical principles* (Vol. V). Oxford, UK: Clarendon Press.

Murray, C. (1997). *What it means to be a libertarian.* New York: Broadway Books.

Nagel, E. (1961). *The structure of science.* NY: Harcourt, Brace & World.

Nagi, S. Z. (1965). Some conceptual issues in disability and rehabilitation. In M. B. Sussman (Ed.), *Sociology and rehabilitation* (pp. 100–113). Washington, DC: American Sociological Association.

Nagi, S. Z. (1969). *Disability and rehabilitation.* Columbus, OH: Ohio State University Press.

Nagi, S. Z. (1991). Disability concepts revisited: implications for prevention. In A. M. Pope & A. R. Tarlov (Eds.), *Disability in America: Toward a national agenda for prevention* (pp. 309–327). Washington, DC: National Academy Press.

NCHS (2007). *Health, United States 2007.* Hyattsville, MD: National Center for Health Statistics.

Neander, K. (1991). The teleological notion of function. *Australasian Journal of Philosophy, 69*, 454–468.

Nissen, L. (1997). *Teleological language in the life sciences.* Lanham, MD: Rowman and Littlefield.

Nordenfelt, L. (1987). *On the nature of health.* Dordrecht: Reidel.

Nordenfelt, L. (1997). The importance of a disability/handicap distinction. *Journal of Medicine and Philosophy, 22*, 607–622.

Rahdert, M. C. (2000). Arline's ghost: some notes on working as a major life activity under the ADA. *Temple Policy & Civil Rights Law Review, 9*, 303–331.

Reznek, L. (1987). *The nature of disease.* London: Routledge and Kegan Paul.

Reznek, L. (1997). *Evil or Ill?* London: Routledge and Kegan Paul.

Rondinelli, R. D., & Duncan, P. W. (2000). The concepts of impairment and disability. In R. D. Rondinelli & R. T. Katz (Eds.), *Impairment rating and disability evaluation* (pp. 17–33). Philadelphia: W.B. Saunders.

Taber, C. W. (1940). *Taber's cyclopedic medical dictionary* (1st ed.). Philadelphia: F. A. Davis. (20th ed., 2005).

Taylor, N. B., & Taylor, A. E. (Eds.). (1953). *Stedman's medical dictionary* (18th ed.). Baltimore: Williams and Wilkins. (5th ed., 2005).

Thomson, J. (1971). A defense of abortion. *Philosophy and Public Affairs, 1*, 47–66.

Tucker, B. P., & Milani, A. A. (2004). *Federal disability law in a nutshell* (3rd ed.). St. Paul, MN: West.

UPIAS (1976). *Fundamental principles of disability.* London: Union of the Physically Impaired Against Segregation.

Wakefield, J. (1992). The concept of mental disorder: On the boundary between biological facts and social values. *American Psychologist, 47*, 373–388.

Wakefield, J. (1999). Evolutionary versus prototype analyses of the concept of disorder. *Journal of Abnormal Psychology, 108*, 374–399.

Wasserman, D. (2000). Stigma without impairment: Demedicalizing disability discrimination. In L. P. Francis & A. Silvers (Eds.), *Americans with disabilities* (pp. 146–162). New York: Routledge.

Wasserman, D. (2001). Philosophical issues in the definition and social response to disability. In G. L. Albrecht, K. D. Seelman, & M. Bury (Eds.), *Handbook of disability studies* (pp. 219–251). Thousand Oaks, CA: Sage.

Weber, M. C. (2007). *Understanding disability law*. Newark, NJ: LexisNexis.

Whitfield, G. (1997). *The disability discrimination act: Analysis of data from an omnibus survey*. London: Department of Social Security.

WHO (1980). *International Classification of Impairments, Disabilities and Handicaps* [ICIDH]. Geneva: World Health Organization.

WHO (1994). *International Statistical Classification of Diseases and Related Health Problems*, tenth revision [ICD-10]. Geneva: World Health Organization.

WHO (2001). *International Classification of Functioning, Disability and Health* [ICF]. Geneva: World Health Organization.

Wouters, A. (2005). The function debate in philosophy. *Acta Biotheoretica, 53*, 123–151.

Wright, L. (1973). Functions. *Philosophical Review, 82*, 139–168.

Zimmer, M. J., Sullivan, C. A., & White, R. H. (Eds.). (2008). *Cases and materials on employment discrimination* (7th ed.). New York: Aspen.

Part II
Disability, Quality of Life, and Bioethics

Chapter 5
Utilitarianism, Disability, and Society

Torbjörn Tännsjö

5.1 Introduction

What are the social implications of utilitarianism with respect to disability? In this chapter I give a rough answer to this question: the implications are well in accordance with our common sense thinking, once various prejudices in relation to disability have been exposed. In order to be able to establish this claim, there are some obvious preliminaries that must be sorted out. Before I can develop my argument, I must say something about what I mean by "utilitarianism" and "disability," and also about how, in the present context, I conceive of a "social" implication. I turn first to the notion of disability, then move on to utilitarianism, and finally to various possible social responses to disability.

5.2 The Meaning of the Word "Disability"

What do we mean when we speak of "disabilities"? I suppose different people mean different things. When I try to search for a definition on the Internet, I find 23 different suggestions.[1] In the U.S., the Americans with Disabilities Act defines a disabled person as someone who has "... a physical or mental impairment that substantially limits one or more of the major life activities of such individual."[2] There is no such thing as the correct use of the word, however. Hence I will make no attempt to find and specify such a use. Instead, I will be happy to make a stipulative definition, intended to suit the purposes of the present paper. This definition is much more inclusive than the one just quoted from the Americans with Disabilities Act. And still it allows us to distinguish between disability and disease. I will say that a person is disabled, in a certain respect, if, and only if, the person falls below what is a species normal variation in the respect in question.

T. Tännsjö (✉)
Department of Philosophy, Stockholm University, Stockholm, Sweden
e-mail: torbjorn.tannsjo@philosophy.su.se

D.C. Ralston, J. Ho (eds.), *Philosophical Reflections on Disability*, Philosophy and Medicine 104, DOI 10.1007/978-90-481-2477-0_5,
© Springer Science+Business Media B.V. 2010

I take it that disabilities manifest themselves in the form of a lack of a capacity for doing certain things that people in general are capable of doing. Disabilities may be absolute in nature. If I suffer from infertility, it may the case that, whatever I do, I cannot conceive children. Or, a disability may appear in a more relative form. If I am mentally retarded, I may be able to perform some intellectual tasks, but not others, and not as easily as do people in general.

On this understanding of a disability, it relates crucially to things we *do*. This is why it is possible to distinguish disability from disease. Being in pain, furthermore, is not to suffer from a disability. However, severe pain, as with many diseases, may be the source of various disabilities. If I am constantly in severe pain it may be impossible for me to focus upon and to solve certain intellectual problems, for example. And if I suffer from a serious disease, I may be too weak to perform tasks most people can perform easily.

Moreover, and most importantly, on this understanding of disability, there is no implication that a disabled person must have a low quality of life. Disabilities may hamper our capacity to lead a happy life, but they need not do so.[3]

As a matter of fact, on this notion of disability, many disabilities go unnoticed. This is true of dyslexia in an illiterate society. There are probably many examples we do not even think about. It is likely that each person is disabled in some respect.

I admit that the notion here defined is both vague and somewhat conventional. When does a person suffer from mental retardation to such an extent that we may correctly classify it as a disability? The correct answer to this question happens to be: when that person falls below a score of 70 on an IQ test. What is so significant about an IQ score of 70? Nothing at all—this is just where we have decided to draw the line. The same is true of other sharp distinctions in the field. We could have drawn them otherwise. If we know that they are conventional, however, and if we realize that they are there in order to help us to handle a vague notion, there is nothing problematic about their being conventional.

5.3 Utilitarianism

This is a study in applied ethics. The ethical theory I want to apply to problems related to disability is classical hedonistic utilitarianism. I will not go into details when it comes to an explanation of utilitarianism. Moreover, I will not try to defend my belief that classical hedonistic utilitarianism is the moral theory that comes closest to the truth. My interest is to inquire as to its implications in this restricted field. And the result I arrive at, that utilitarianism seems to have quite acceptable consequences in the field, should not be interpreted automatically as evidence in support of it. Other views may have similar implications. Our adoption of one view in particular, among them, must rest on an assessment of which one offers the best moral explanation of the "data." In the present context, I will not delve into this discussion.

Something should be said about the content of the view I want to apply, however. It should suffice to say here that the theory I discuss claims that we have an obligation always to act so that we maximize the sum-total of well-being (happiness) in the world. This, then, is the *criterion* of rightness given by utilitarianism. This criterion presupposes that it is meaningful to make both intra- and interpersonal comparisons of happiness. When I say that one person is happier than another person, right now, there is a truth of the matter about that claim. I also assume that it is possible, in practice, to make at least rough estimates as to how happy or unhappy people are. This renders it possible for us to construct a *decision-method*, informed by the utilitarian criterion of right action. Roughly speaking, according to the decision-method in question, when we construct social institutions we ought to do our best to maximize *expected* happiness. The rationale behind the adoption of this decision-method is the belief that, if we stick consistently to it, we will achieve a better result than if we were to adopt any other method we can think of. This means, of course that when we try to apply utilitarianism, we have to rely on many simplifications of the problem at hand.

Moreover, considering the utilitarian criterion of right action, it should be pointed out that there are two ways of maximizing the sum-total of happiness in the world: either by making existing individuals happier, or by making happy (additional) individuals.[4]

My query in the present paper, then, concerns the implications of this theory for how society ought to respond to disabilities.

Part of my query is how a publicly-subsidized system of health care should respond to disabilities, given utilitarianism. I assume that utilitarian concerns support the establishment of a publicly financed health care system in the first place. This is clearly not the place to defend such a presupposition—here, I merely take it for granted.

I also discuss how society should respond in other ways to disability, for example by compensating disabled people for incurable disabilities, as well as what kind of approach society should take up with respect to techniques of prenatal genetic testing and selection against fetuses and embryos with a disposition towards disability.

5.4 Different Forms of Disabilities

When disability is defined as an aberration from what is a species typical variation with respect to some kind of capacity for action, it is possible to discuss to what extent disabilities pose a threat to the subjective well-being (happiness) of people. Since I have not defined the notion of a disability with any reference to the hedonistic status of disabled people, it must be an open empirical question to what extent disabilities make people unhappy. Here different possibilities seem to exist, and it is useful to keep them distinct.

(a) Some disabilities entail no loss of happiness whatever. I will speak of such a disability as a "mere" disability. Clearly, if disability is seen merely as a lack of capacity for doing things that other members of the human species are typically capable of doing, then, in many cases, a disability should pose no threat whatever to the hedonic state of the disabled person. I have already remarked that many disabilities simply go unnoticed. This is true of dyslexia in an illiterate society, and there may exist many other examples. It is likely that each of us is disabled in some respect, without our knowing about it. But it may even be true of many disabilities that we notice that, typically and as such, they do not affect the hedonic status of the disabled person. Being tone-deaf or color-blind is hardly discussed as a disability at all, but on my notion of a disability they are both examples of disabilities; probably, tone-deaf and color-blind people lead as happy lives as do people with a perfect pitch sensitivity and color vision. More interestingly, infertility seems to be a mere disability. This flies in the face of received wisdom, but there is no consistent support in happiness studies for the claim that couples with children are any happier than couples without children. Empirical evidence suggests that, even if having children in general improves the quality of life, the effect is of an amazingly short duration. Within two years parents revert on average to their original level of happiness.[5] There is little evidence to the effect that the long-term consequences of having children are significant. Some studies show that children make no difference to the happiness of the life of the couple.[6] Some studies indicate that married couples with children are *less* happy than married couples without children.[7]

(b) Other forms of disability do pose a threat to the happiness of the disabled person. And yet, the threat may be easily averted. I will speak here of "simple" disabilities. I think of such states as blindness and deafness and mental retardation. Here are some typical reports about these disabilities. Of course, with respect to all these findings it is necessary to keep in mind that diseases causing the disability in question, as well as the prejudiced reaction from society to it, may compromise the happiness of the disabled person. In these studies, attempts have been made to compensate for such factors. First, a study regarding blindness conducted by Kleinschmidt et al.:

> ... depression, anxiety, and decreased life satisfaction are not necessarily long term consequences of vision loss. After an initial reaction phase, they will not be experienced by the majority of the visually impaired.[8]

Second, a study concerning deafness notes that, not even among old people with pre-lingual deafness, who experience a greater number of ill-health symptoms such as depression and insomnia, is reported perceived subjective well-being compromised:

> The results strengthened the assumption that depressive symptoms and sleep disturbance are more frequent among elderly pre-lingually deaf people using sign language than among hearing people. On the other hand, and contrary to our expectations, this did not imply significantly lower perceived subjective wellbeing compared with hearing elderly people.[9]

Finally, with respect to intellectual disabilities (ID), there are even reports to the effect that people "suffering" from them lead more happy lives than people who don't.[10]

As such, then, these simple disabilities do not seem to make people less happy than people who do not suffer from them. However, in order to remain happy while blind or deaf or mentally retarded, one needs all sorts of measures to be taken by society in order to facilitate one's life—for example, traffic lights combined with

sound for the blind, special education in sign language for deaf people, and a special
kind of nursing home for severely mentally retarded people. Still, given appropriate
adjustments from society, if one suffers from a simple disability, that person is likely
to live his or her life just as happily as do other people who do not have the disability
in question.

With respect to simple disabilities we should distinguish between two kinds. First
of all, we have simple disabilities that are likely to affect most of as during our life
cycle. They are most efficiently taken care of by very general measures, rendering
society accessible to everyone. This is hardly controversial. It is in the interest of us
all that society adapt to the existence of people with these disabilities and provide
necessary services and adjustments, since it is likely that most of us, at some stage
of our lives, will become dependent on them. In the individual case, then, such a
simple disability is not associated with any social cost (i.e. any cost to others).

These measures should be distinguished, secondly, from more specific measures
directed at those who are in need of them, where it is not in any direct interest
of society at large (people in general) that these measures be taken. Society can
still cover these costs, of course. Another possible reaction is not to do so but to
insist instead that disabled people pay for their own services. I suppose that in
a publicly-financed healthcare system, established for utilitarian reasons, only the
former option is reasonable.

(c) A third category of disabilities does indeed have the effect that those who
suffer from them are affected negatively with respect to their hedonic status. There
may be ways of rendering the lives of these people easier, and there may be ways of
compensating them for losses they suffer because of their disabilities, but here the
disabilities as such do rob the disabled individuals of some happiness that is available to people who are not in this respect disabled. I will speak of such disabilities
as "problematic" disabilities.

Are there examples of problematic disabilities? On a standard, unreflective
understanding of disability, most disabilities fall in this category. However, in fact
they may be much less common that we usually think. We think they are common because we are the victims of prejudice with respect to these disabilities. And
the reason that these problematic disabilities are rare—if they are—lies, then, in a
general human capacity to adapt to loss of abilities.[11]

Still, some problematic disabilities do exist. Here, if we are to trust happiness
studies, are some examples. Spinal Cord Damage (SCD) does not mean that your
life is not worth living, but it does mean that your life, even when you have adapted
to the disability, is likely to be lived on a slightly lower level of subjective well-
being than the one felt on average by people lacking such damage. This is how the
situation of paraplegics is characterized by one important researcher in the field,
Marcel Dijkers, summing up the relevant literature on the topic:

> . . . subjective well-being among persons with SCI was lower than in the population at large,
> but . . . the difference was not dramatic.[12]

It should be noted that while being blind or deaf means no loss as such of happiness,
being hard of hearing does seem to contribute to a less happy life than the lives

people live on average. Being hard of hearing is a problematic disability, then. Here is a typical assessment:

> The hard of hearing have worse social relationships than the signing deaf, and are disadvantaged relative to the hearing in all areas measured. Quality of life is related to the level of satisfaction with the hearing achieved by hearing aids.[13]

There are of course many more examples that could be mentioned. And, since such problematic disabilities do exist, the question can be raised how society should react to them.

(d) Finally, there is a category of disabilities that are so severe that, whatever we (science, medicine, and so forth) do, they rob the disabled person of a life worth living. These disabilities are sufficient, all by themselves, to rob the disabled person of any chance of experiencing positive happiness. He or she is destined to remain, for the rest of life, below the level where life is positively worth experiencing. I will speak of such disabilities as "tragic" disabilities.

Are there really examples of tragic disabilities? Here again we must guard ourselves against prejudice. I saw, for example, in a recent European survey that one third of all Turks believe it is better to be dead than to be sexually impotent.[14] In this they are most likely wrong. How can they be so wrong? Well, most of them have no experience of being sexually impotent, either because they are women, or because they are men and sexually potent. We should not ask these people about what it is like to be sexually impotent. We should ask people with experience of a disability what it is like to suffer from it, not people who have no experience of it. And we should be careful only to ask people who have had a chance to adapt to their situation. My conjecture, then, is that most sexually impotent people have been able to adjust to their disability. We should bear in mind that Leonard Cohen, the singer and songwriter, when cautioned that anti-depressant medication may mean a loss of sexual appetite, answered that, on his understanding, this was part of the very point of taking them! Still, I think there are some rare examples of tragic disabilities. One clear example would be a terminal disease such as ALS, which progressively robs a person of her capacity to breathe; eventually she will suffocate. In late stages of such a disease, the disability caused by the disease may well be such that it would be better not to have to experience life any more; in such circumstances, it would be better to be dead or sedated into oblivion.

How should society respond to such disabilities?

5.5 A Caution About the Meaning of Life

It should be stressed that the version of utilitarianism here taken as a point of departure, classical hedonistic utilitarianism, gives a rather special—and controversial—answer to when a life is going well for the person living it. This is entirely a matter of how much experienced well-being that life contains. I assume this as a matter of fact. This has nothing to do with whether intrinsic preferences have been satisfied, or with whether important achievements have been accomplished. A perfect life may be a life with very weak preferences, and most of them frustrated, and with hardly any achievements, but where the person living the life in question is constantly quite happy. Most of the time, the person living the life in question, whether awake or dreaming, has been well above the threshold where life begins to be worth experiencing. We may speak here of narrow hedonism. I assume that narrow hedonism is correct.

I make this assumption because I believe that narrow hedonism does indeed give the true answer to the question about the meaning of our lives. However, I will make no attempt to defend this claim.[15] It is important, though, that it be made explicit, since much of what I have to say about what are the reasonable social reactions to disabilities hinges on it.

For example, on a perfectionist view of the good life, what I say about mental retardation may seem wrong. If you believe that you do not lead a good life unless you solve intricate intellectual problems, then you are not likely to accept the kind of argument I put forward in the present context. Not even a refined, or qualified, hedonism of the kind defended by J.S. Mill, will lead to the results at which I arrive. After all, according to Mill (1863/1962) it is better to be a dissatisfied Socrates than a satisfied fool (a mentally retarded person, as we would say today). I do not share his conviction, and in the present context I will just assume that he is wrong.

On the face of it, it may seem as though many happiness studies, or studies of *subjective well-being* (in contrast to studies measuring life *satisfaction*), measure happiness or hedonic status as defined here. Appearances may be deceptive, however. I must caution the reader that I find much of the research in this field problematic. The problem of a hedonistic unit, a theme in the philosophy of the classical hedonistic utilitarians Bentham and Edgeworth, is not taken seriously by modern happiness studies.[16] It is difficult to understand how these modern scholars can be so certain that it is possible to compare intervals of happiness between people, in the way they presuppose that they do. This kind of comparison seems to be difficult to achieve even with simple assessments of pain. I happen to have had an illness (Ileus), which brought me to the point where I lost consciousness because of intense pain. Since then I know where my upper limit is. I can hence place more mundane pains on a scale where 0 is no noticeable pain at all, and 10 is the point where I faint. However, it is an open question whether my upper limit is the same as the upper limit of other people.

Furthermore, even if I can say at some time that my feeling of pain should rate at 5, this does not give much information about my hedonic state. A person who is giving birth to a wanted child, and suffers pain at 5, may be extremely happy. Another person, suffering from a terminal disease, giving rise to pain at 5, may be way below the line where life is worth experiencing. I know of no scholar in the field of happiness studies who has taken this problem seriously.[17]

Does all this mean that we know nothing at all about the relation between disability and happiness? Well, it does mean that our evidence is very problematic. And this means that my classification of disabilities as mere disabilities, simple disabilities, problematic disabilities, and tragic disabilities, is extremely speculative. The classification may in some cases be altogether wrong. However, since the results of happiness studies is the only thing I as a philosopher can go on when discussing the problem, I have just assumed that they are, roughly speaking, correct. If they are not, many of my conclusions would be compromised. And yet, what is wrong, if these studies are wrong, is not the general analytical tools here devised, but merely the classification of certain disabilities.

I hope further research may help us to shed more light on this problem. But, in order for this to happen, the scholars performing happiness studies must start to take seriously the neglected problem of the hedonic unit.[18]

5.6 How Should Society Respond to Disabilities?

I am now in a position to raise the question how society should respond with respect to disabilities. However, it is necessary, even here, to make some further distinctions. I have already hinted at them. One question has to do with how society should distribute scarce medical resources when it is possible, through medical interventions, to cure disabilities. Another question has to do with the possibility of compensating disabled people for losses in well-being, when no cure is possible. A third question has to do with the possibility of saving resources for society at large, by preventing disabilities.

The first of these questions, concerning a cure, can be raised meaningfully, as we will see, with respect to all kinds of disabilities: mere disabilities, simple disabilities, problematic disabilities, and tragic disabilities. The second question, concerning compensation, can be raised meaningfully with respect to simple disabilities and problematic disabilities, while the third question, concerning selection, can be raised meaningfully, once again, with respect to all kinds of disabilities.

In particular, when it comes to prevention, it is worthwhile to discuss whether society ought to save resources, and to create a situation where the lives lived are free of simple but costly, problematic, and tragic disabilities, by *preventing* the birth of people with those simple but costly, problematic, or tragic disabilities. We can also discuss whether society should allow parents, if they wish, to select against mere disabilities.

I will address these questions in order.

5.7 Disabilities and the Allocation of Scarce Medical Resources

In ordinary circumstances, as a rule of thumb, given a publicly-financed system of healthcare, utilitarians trying to maximize the sum-total of expected happiness tend to argue that medical resources should be used where they most efficiently produce happiness in the person submitted to the treatment. I will conduct my discussion under this assumption; this does not mean that there are not cases where, when we distribute scarce medical resources, we must keep an eye also on the supply side of the equation. This is true in situations of mass casualties, pandemics, and so forth, where it is crucial first of all to save those people who can in their turn save other people, or keep the basic structure of society going (such as fire brigades, paramedics, the police, and the government).[19] However, such situations pose different questions, and I will simply set them to one side. I will also abstract from possible indirect effects of medical interventions on those who are near and dear. My focus here is on the question: to what extent do we make the best use of scarce medical resources when we direct them to the curing of disabilities—where the benefit is assessed *exclusively* from the point of view of the person who undergoes the cure? This is a simplification, but it renders more comprehensible and easily assessable the main thrust of my argument.

I have remarked that there are disabilities that go unnoticed or, even if they are noticed, they do not as such mean a threat to the well-being of the disabled person. I have given as examples color blindness, tone deafness, and infertility. I have chosen to speak here of *mere* disabilities. Does a utilitarian approach to mere disabilities mean that nothing should be done to cure them, at least when these cases are assessed solely from the point of view of the putative patient?

This may seem to be the obvious answer, but here a rather strange complication exists. It may well be true that a person who is tone deaf can lead a life as good as a person with a perfect pitch, and that a person who is infertile—and hence cannot conceive biological children of his or her own—will live, on average, as happily as a person who can, and who does, conceive children. However, once a cure exists for tone deafness or infertility, it is likely that many people will ask for it. And if it is denied them, it is highly probable that this will cause not only noticed frustration of a desire, but also felt unhappiness.

This, then, is a perverse consequence of medical development. We invest scarce resources in cures. So long as there is no hope of a cure, there is no need for it. People adapt to their disability (their lack of a capacity that other people have). Once it becomes possible to have this capacity, they come to want it, ask for it, and become unhappy if they are not provided with it. So, perhaps it is mandatory for society to provide it. Now, however, perversity transforms into futility.[20] Those who receive the service revert to the level of happiness where they started, and where they would have remained, had no one thought of any cure.

Even the *search* for a cure may have the perverse effect that people come to see their disability as a problem and the solution to the problem as a right. More resources must be spent on research, on new technologies, and eventually on costly cures. And the net result is that we are back where we started.

Moreover, when a certain measure is taken because it is needed by some people—such as cochlear implants for those who are hard of hearing, and who really need them in order to maintain their happiness—deaf people, who do not need them to maintain their happiness, may come to request them for themselves or their deaf children as well, and become utterly frustrated and unhappy if they do not receive them too.

The utilitarian solution to this perversity/futility problem is, I submit, to surrender to it. At least we have to acknowledge that, once the cure exists, it should be provided to all who make a request for it. It is difficult to limit scientific development only to possible cures for mere disabilities. Hence we are stuck with the perversity/futility problem.

What, then, about problematic disabilities? A typical example would be spinal cord injuries. Being paraplegic due to an acquired spinal cord injury means, as we have seen, some loss of happiness, but not much. This may come as a surprise to those who lack familiarity with the literature in the field and who have no personal experience of the situation of people in this category. Yet, this seems to be the consistent result of all studies in the field. Assume that this empirical view is correct; where does this leave these people in the competition for scarce medical resources?

In an affluent society, this means that they have a robust claim for the resources in question—unless it is extremely expensive to cure their disability. We can speculate here. Suppose the loss of happiness when you become a paraplegic is 10%, on average. This means that, each day, on average, you experience only 90% of the happiness experienced by a person who can walk. If we look at an entire life with such a problematic disability, then this means something comparable to an untimely loss of 8 years of your life—if we suppose that an ordinary life goes on for some 80 years. So a person with a problematic disability of this order can compete successfully for medical resources that alleviate his or her problems on a par with a person in his or her 70s who suffers from a life-threatening condition.

Even though I do not intend to defend hedonistic utilitarianism against competing views, it might be interesting nevertheless to know whether its assessment here is any different from that of egalitarian thought. In a brief aside, let me briefly give the answer to this question.

5.8 Egalitarianism—An Aside

What is called egalitarian thought is really a mixture of many views. Here, I will focus only on the two most plausible ideas in the field. They can be characterized as follows.

One view is egalitarianism proper. This is the view that we should try to level out happiness between people—even at some cost, considered from the point of view of the sum-total of happiness. The rationale behind the view is the idea that we lead separate lives. In its most plausible version, egalitarianism proper looks at entire lives.[21] Furthermore, in its most plausible version, egalitarianism places value on equality in such a manner that the view is consistent with a requirement of Pareto optimality. A change which means that some lose well-being, while no one gains any well-being, is never advocated by this version of the theory.[22]

The other view is an idea to the effect that there is a diminishing marginal moral importance of happiness. The less happy you are, the more important is a certain increase in happiness. In particular, it is more important to alleviate severe suffering than to increase intense happiness.

This latter view is usually called prioritarianism. According to prioritarianism, in its most plausible interpretation, the primary interest is not in entire lives but in moments.[23]

The priority view can be seen as a revision of utilitarianism. However, the view can also be combined with egalitarianism proper. You must then conclude that, if one individual, A, has lived constantly at +10, hedonically speaking (for 80 years, say), while another person, B lived his first 40 years at −100 and his last 40 years at +120, the former person, A, may have lived a better life, on the whole, than the latter one, B. The latter person, B, is worse off, then, and should be given an additional benefit, rather than the former person, A, if egalitarianism proper is taken as our point of departure for moral assessments. The time at -100 weighs heavily, while the time at +120 weighs lightly, when assessed from the point of view of prioritarianism.

I have no sympathy for either egalitarianism proper or the priority view, or their combination, but I will still say something about their implications with respect to disability, in order to put the utilitarian answer in some perspective.

Is egalitarianism proper, or the priority view, more demanding on us all (society) than utilitarianism, with respect to problematic disabilities?

I think not. A disabled person, in this category, is not among those who are worst off, when we look on the matter from the point of view of egalitarianism proper. His or her life compares to a person who can live to his or her 70s. And, since the disability is merely problematic, not tragic, the loss of happiness it causes as such does not count among those the cure of which must be prioritised.

5.9 Back to Utilitarianism

What if there is no cure for the problematic disability in question? Should the disabled person still be compensated? Does the disabled person have a right. say, to personal assistance, comparable to the right he or she would have had to a cure, had a cure existed? I return to utilitarianism when I try to deal with this question. And the correct answer seems to be in the affirmative. The person with a problematic disability, who cannot be cured, should have compensation of some sort. And the most obvious compensation is personal assistance of some sort, thereby rendering life easier.

Does this set a problematic precedent? What if a sick person, who is offered treatment within a publicly-financed health care system, claims that he or she is prepared to abstain from the cure society offers, if he or she is instead given a lump sum equivalent to the cost for the treatment? I suppose we would not like to accept this. But if we compensate people with problematic disabilities, why should we not accept this claim by the patient in question?

One possible answer might be that there is a risk that this patient will change his mind and return later with a new request for treatment. It would be difficult for society to refuse this treatment, which can easily be conceived of as a treatment for another (not exactly the same) condition.

What if a patient suffers from a disease that causes no disability, but that is painful for the patient, and is incurable? Does this mean that this patient should be (economically) compensated? My intuition is that he should not be compensated. But can we stick to this answer, if we allow that people with problematic disabilities be compensated?

Why is compensation for suffering caused by disability special? Perhaps my intuition is misleading. Perhaps people with all sorts of suffering, not possible to cure, should be offered reasonable compensation, if such is possible.

It strikes me as more reasonable to provide personal service to a person who is a paraplegic than to offer money to a person who suffers from refractory pain, but I find it difficult to provide any rationale for this intuition, so perhaps it is just wrong. And, certainly, if compensating those with problematic disabilities means that, on pain of inconsistency, we must also compensate people with refractory pains, then I think we should compensate people with refractory pains, even economically, if there is no other way of doing it.

5.10 Tragic Disabilities

If we turn finally to tragic disabilities, the situation is different. I have here assumed that there is no cure. Should one surface, then of course it should be given to people in this category. This would have high priority on both utilitarianism and the priority view. It would have less urgency on egalitarianism, though, at least if it has a late onset in a long life. This counts against the plausibility of egalitarianism proper, as far as I can see.

I have defined the notion of a tragic disability such that no compensation is possible for the tragically disabled person. What, then, if no cure exists; is there nothing that the health care system has to offer to the disabled person? Of course, there are many things the health care system could still offer.

First of all, the tragically disabled person should be given all sorts of palliation. Moreover, physician assisted suicide and euthanasia are services society should offer to him. It would be cruel, and a waste of resources, not to provide them.

This is even more obvious, given a utilitarianism revised in the light of prioritarianism, of course. And yet, since these patients are already given high priority on standard utilitarian reasoning, the practical importance of the theoretical difference between utilitarianism and prioritarianism in this respect is not very significant.

Again we find, somewhat surprisingly, that according to egalitarianism, at least if it has not been tempered with some prioritarianism, the needs of very old people in terminal anguish carry relatively little weight. They are, after all, among the winners in life, not among those who are worst off.

5.11 A Special Consideration

I have indicated that, somewhat surprisingly, infertility is a mere disability. Through the perversity/futility-mechanism, in vitro fertilization (IVF) treatment may still gain some utilitarian support. Since it exists, people make requests for it, and if their requests are turned down, they will not only become frustrated, but unhappy as well.

Is this the end of the matter? It is not. On classical total utilitarianism, there exists a special argument why we should give very high priority to IVF treatment for infertile couples. The reason we should do so is that there are two ways of making the world a better place. One is by making existing people happier. The other is by making happy people. When IVF is provided to infertile couples we often do the latter. The beneficiaries of this treatment are the children brought into existence.

It should be noted that the cost effectiveness of an IVF treatment is extremely high: at a moderate cost we create some 80 expected years of perfect, healthy and happy life. Few medical interventions can compete with this kind of achievement. Infertility treatment is in a class of its own, in the competition for scarce medical resources.[24]

5.12 The Problem of Selection

Thus far I have focused on the patient. I have asked whether we should cure a person with a disability, and I have asked whether we should compensate him or her, when no cure exists. When scarce resources are discussed, it is also of interest to consider whether we can somehow affect the scope of the problem. Is there a way of allocating more resources to those who need them, by keeping the number of needy persons down? And are there reasons to see to it that those people who live are also, among all possible people, those who will live the most happy lives?

The answer to these questions, in the abstract, if we answer them with reference to classical hedonistic utilitarianism, is simple and in the affirmative. Other things being equal, we ought to save resources. Other things equal, it is better if very happy people live rather than less happy ones do. Even if there is no one there to experience any difference, we ought to opt for the world where the sum-total of happiness is maximized. And yet, medical measures such as prenatal diagnosis and abortion in general, and Preimplantation Genetic Diagnosis (PGD) in particular, tend to give raise to extremely sensitive questions. Should they really be provided? On what terms should they be provided, in that case?

As I have indicated, the answer to these questions, when arrived at from the point of view of utilitarianism, seems obvious. We avoid unnecessary costs, if we resort to these methods—most obviously so in relation to tragic and problematic disabilities, but even in relation to simple disabilities as well. Even mere disabilities may cause unnecessary costs (because of the perversity/futility-mechanism). And not only is this a question of costs to society; many disabilities mean a loss of happiness since, had other people lived instead of the disabled people, they would have been happier (which is certainly true of tragic and problematic disabilities).

If I come to think, therefore, as an individual, about the problem, I must, given that I am a utilitarian, come to the conclusion that, if I can avoid the risk of having a problematically or tragically disabled child, I ought to try to avoid this. It may even be the case that I ought to try to avoid having a child with a simple disability, if this disability means an unnecessary cost to society. Even a mere disability may be a reasonable target for PGD, since there is always a risk that my child will, perversely, want to get rid of it.

Does all this mean that prenatal diagnosis, and in some cases, PGD, should be obligatory measures for responsible parents to undergo? Should selective abortion be a legal obligation?

I think not, and the reason is that such a policy would mean a felt and very serious threat to people living with disabilities of various different kinds. The mere exercise by some of a right to choose children means that people living with various disabilities may come to feel that their lives are put at risk. This is a concern often raised by representatives of organisations of disabled people. And this concern is a serious and a genuine one. It should be taken seriously. However, and alas, there is no ideal way of handling it.

Of course, if we could prohibit all kinds of use of techniques rendering a selection of children to be born possible, we may have obviated the threat felt by people living with diseases and conditions selected against. However, this is neither a reasonable nor a feasible strategy. It is unreasonable because it requires, in rare cases, that responsible putative parents are not allowed to behave in a moral manner; instead they are forced into risking the birth of children who are destined to lives not worth living (i.e. lives with tragic disabilities). And it is not feasible since these techniques are already with us, and a prohibition in one country would only lead to medical tourism.

How then, can we best handle the concern, if not by prohibiting the selective techniques? Should we regulate their use or should we permit them to be used as prospective parents see fit? These seem to be the two remaining options.

It is tempting to argue that, while some choices should be allowed, others should be prohibited. Prospective parents should not be allowed to choose the sex of their children, for example–let alone, if it becomes possible, their children's sexual orientation. Why? Well, I suppose the argument must be that it is no better to be born with one sex rather than the other, or with one sexual orientation rather than the other.

However, if this is how society regulates the use of the selection techniques, it sends out a message. Partly, this message is fine: there is no problem being female or gay. However, when at the same time society allows for other reproductive choices, such as a choice against a child with a mental disability (Down syndrome, say), it does send out a rather nasty message. Down Syndrome is indeed a problem! This means that the Nazi spectre is once again alive.

So in order to avoid a situation where society has a view on what kind of lives are worth living, we should allow prospective parents to exercise complete freedom in this respect.[25] This means that, while some parents find, for example, that a deaf child is too much right now (they are not prepared to migrate into another culture, which is alien to them, they are not prepared to learn a new language, and so forth), these parents are allowed to make a selection against deafness. At the same time, another (deaf) couple may welcome a deaf child; they are even free to make a selection in favor of deafness.[26]

Does this mean that many children will be born because of immoral choices made by their parents? Does it mean that some children will be born to lives not worth living? Will it mean that some parents deliberately conceive children with fewer chances of having full and happy lives than children with better chances?

To some extent, this is bound to happen. However, in most cases prospective parents are very eager to see to their prospective children's best interests. It is more likely that parents will make idle choices, such as choices against mere disabilities, rather than foolish choices, in favor of problematic disabilities. Yet such choices, even if idle, are innocuous. You don't live a *worse* life when you lack a mere disability than you would have lived, had you had this disability. It makes no difference whether you have it or not. So in most cases, procreative freedom and a freedom to select will have good consequences, or consequences without any significance whatever.

Moreover, there is no guarantee that society will be successful if, instead, it takes over responsibility for these choices. And when society goes wrong, it may go wrong on a large and terrifying scale.

Finally, even if society (the state, the doctors, the medical authorities, or what have you) were more successful than most prospective parents in making these decisions, which I very much doubt that it would be, the very fact that we had endowed society with the right to decide what sort of people there should be would mean a serious threat to disabled people. So we had better resist any temptation to adopt such eugenic policies.

5.13 Conclusion

I have tried to find out what the social implications of classical hedonistic utilitarianism are with respect to different kinds of disability: mere disability, simple disability, problematic disability, and tragic disability. The implications seem to be, once we rid ourselves of prejudice with respect to disability, rather commonplace. Very roughly, the implications are as follows.

First of all, there are reasons, within a publicly-financed health care system, to try to cure disabilities. This is true even of some mere disabilities, where people very much want to get rid of them—in spite of the fact that, as such, they do not mean any threat to their level of felt happiness. Here society must surrender to the perversity/futility-mechanism identified in this paper. People with disabilities have a rather strong claim on scarce medical resources, given utilitarianism. Somewhat unexpectedly, they have, in many cases, an even stronger claim than the one they would have had, if instead some kind of egalitarianism or prioritarianism had been the point of departure of the assessment of their needs.

Secondly, when people suffering from problematic disabilities cannot be cured, these people have a right to compensation from society, once a publicly financed health care system has been established. The need for compensation, mainly in the form of personal assistance, is no less urgent than the need for a cure, when a cure exists. Somewhat unexpectedly, it has also been concluded that people suffering from intractable pain (but no disability) have a similar right to compensation.

Finally, even if there are good utilitarian reasons, in the individual case, to avoid the birth of people with costly simple disabilities, as well as people with problematic and tragic disabilities, through the use of prenatal genetic diagnosis and selective abortion, as well as through IVF and PGD, it would not be a good idea to make the use of these techniques obligatory. This would be to give rise to speculation among people living with these disabilities that they should not really be where they are; they pose a burden to society. Hence, it is wise utilitarian policy to avoid any concession to eugenics whatever. We may safely assume that, if society is neutral with respect to the question what sort of people there should be, individual couples are capable, in most cases, of arriving at wise answers to this question.

Classical hedonistic utilitarianism seems to give the "right" answer to how society should react to disability, then. This does not mean that we have come across any positive evidence in favor of classical hedonistic utilitarianism. Other moral views, when applied to the same problem, may give rise to similar implications. We have evidence for a moral theory only where it gives the *best* explanation of the data at hand. However, the fact that classical hedonistic utilitarianism gives answers in conformance with our considered moral intuitions does at least show that it has not, in this field, been *dis*confirmed.

Notes

1. See *Definitions of disability on the web* [On-line]. Available at http://www.google.com/search?hl=en&defl=en&q=define:Disability&sa=X&oi=glossary_definition&ct=title
2. See *Americans with disabilities act of 1990* [On-line]. Available at http://www.ada.gov/pubs/ada.htm#Anchor-Sec-47857
3. For a very different, explicitly normative definition, see Kahane and Savulescu (2009). I see no merit whatever in a normative definition; in particular, it would not be helpful to the kind of inquiry I make here.
4. I defend hedonistic utilitarianism in Tännsjö (1998).
5. Clark, Diener, Georgellis, and Lucas (2003).
6. Argyle (2001) and Veenhoven (2004). However, for support of the opposite claim, see De Santis, Seghieri, and Tanturri (2005).
7. Veenhoven (1984) and Argyle (1999).
8. Kleinschmidt et al. (1995, p. 32).
9. Werngren-Elgstrom, Dehlin, and Iwarsson (2003, p. 13).
10. Verri et al. (1999).
11. The adaptation mechanism seems to be very general with respect to adverse events, and it applies in particular to loss of abilities. See Frederick and Loewenstein (1999).
12. Dijkers (1999, p. 867).
13. Fellinger, Holzinger, Gerich, and Goldberg (2007).
14. Unpublished report for a congress on neurosurgery held in Glasgow in 2007.
15. I make an attempt to do so in Tännsjö (1998).
16. See Bentham (1973) and Edgeworth (1881).
17. Even in Kahneman, Wakker, and Sarin (1997), perhaps the most theoretically advanced discussion of the relation between classical hedonistic utilitarianism and empirical happiness studies, the problem is simply glossed over.
18. I discuss, and revise slightly, Bentham's and Edgeworth's idea that the hedonic unit is a just-noticeable difference with respect to well-being in Tännsjö (1998).

19. I discuss such cases in Tännsjö (2007).
20. I borrow the terms from Hirschman (1991), though I use them in a different way.
21. This is a bold claim, of course. It has been denied, for example in Temkin (1993). Yet, I find it very plausible.
22. It is rather obvious that it is possible to give a weight to inequality such that the Pareto condition is satisfied. This has been proved by Wlodek Rabinowicz in a paper in Swedish (Rabinowicz, 2004), generalizing on a proposal in Hirose (2003).
23. This is a bold claim. Once again, however, I can't help finding it very plausible. In many statements of the view it is not clear whether it is understood on a momentary (the most plausible version) or a total life (the least plausible version) interpretation.
24. I will be brief here, since I have been arguing this point over and over again over the years, starting (in English) with Tännsjö (1992) and in Swedish even before that.
25. For an argument against this kind of view, see Buchanan, Brock, Daniels, and Wikler (2001). In my opinion, these authors do not take seriously enough the argument that eugenics is likely to pose a felt threat to disabled people—and they exaggerate the risk that parents, when given total freedom to make their own reproductive choices, will make unwise choices.
26. I defend this view in Tännsjö (1991), Tännsjö (1999), and elsewhere. For a recent defense of such a view, see Savulescu (2002).

References

Argyle, M. (1999). Causes and correlates of happiness. In D. Kahneman, E. Diener, & N. Schwarz (Eds.), *Well-being: The foundations of hedonic psychology* (pp. 353–373). New York: Russell Sage.

Argyle, M. (2001). *The psychology of happiness* (2nd ed.). London: Routledge.

Bentham, J. (1973). Value of a pain or pleasure. In B. Parekh (Ed.), *Bentham's political thought* (pp. 109–127). London: Croom Helm.

Buchanan, A., Brock, D. W., Daniels, N., & Wikler, D. (2001). *From chance to choice: Genetics and justice*. Cambridge: Cambridge University Press.

Clark, A. E., Diener, E., Georgellis, Y., & Lucas, R. (2003). Lags and leads in life satisfaction: A test of the baseline hypothesis. *CNRS and DELTA-Fédération Jourdan* [On-line]. Available at http://www.delta.ens.fr/abstracts/wp200314.pdf

Dijkers, M. P. J. M. (1999). Correlates of life satisfaction among persons with spinal cord injury. *Archives of Physical Medicine and Rehabilitation, 80*, 867–876.

Edgeworth, F. Y. (1881). *Mathematical psychics*. London: Kegan Paul.

Fellinger, J., Holzinger, D., Gerich, J., & Goldberg, D. (2007). Mental distress and quality of life in the hard of hearing. *Acta Psychiatrica Scandinavica, 115*, 243–245.

Frederick, S., & Loewenstein, G. (1999). Hedonic adaptation. In D. Kahneman, E. Diener, & N. Schwarz (Eds.), *Well-being: The foundations of hedonic psychology* (pp. 302–329). New York: Russell Sage Foundation.

Hirose, I. (2003). *Equality, priority, and aggregation*. Ph.D. Dissertation, Department of Moral Philosophy, University of St. Andrews.

Hirschman, A. O. (1991). *The rhetoric of reaction: Perversity, futility, jeopardy*. Cambridge and London: Harvard University Press.

Kahane, G., & Savulescu, J. (2009). 'The welfarist account of disability'. In K. Brownlee & A. Cureton (Eds.), *Disability and disadvantage*. Oxford: Oxford University Press.

Kahneman, D., Wakker, P. P., & Sarin, R. (1997). Back to Bentham? Explorations of experienced utility. *The Quarterly Journal of Economics, 112*, 375–405.

Kleinschmidt, J. J., Trunnell, E. P., Reading, J. C., White, G. L., Richardson, G. E., & Edwards, M. E. (1995). The role of control in depression, anxiety, and life satisfaction among visually impaired older people. *Journal of Health Education, 26*, 26–36.

Mill, J. S. (1863/1962). *Utilitarianism* (M. Warnock, Ed.). London and Glasgow: Collins/Fontana.

Rabinowicz, W. (2004). Om ojämlikhetens negativa värde' ["On the negative value of in equality"]. *Tidskrift för politisk filosofi, 1*, 52–66.

De Santis, G, Seghieri, C., & Tanturri, M. L. (2005). Children and standard of living in old age. Dipartimento di Statistica, Università degli Studi di Firenze, *Working Paper 2005/02*.

Savulescu, J. (2002). Deaf lesbians, "designer disability," and the future of medicine. *British Medical Journal, 325*, 771–773.

Tännsjö, T. (1991). *Välja barn* [To Choose Children]. Stockholm: Sesam.

Tännsjö, T. (1998). *Hedonistic utilitarianism*. Edinburgh: Edinburgh University Press/Columbia University Press.

Tännsjö, T. (1999). *Coercive care: The ethics of choice in health and medicine*. London and New York: Routledge.

Tännsjö, T. (1992). Who are the beneficiaries?' *Bioethics, 6*, 288–296.

Tännsjö, T. (2007). Ethical aspects of triage in mass casualty. *Current Opinion in Anaesthesiology, 20*, 143–146.

Temkin, L. (1993). *Inequality*. Oxford: Oxford University Press.

Veenhoven, R. (1984). *Conditions of happiness*. Dordrecht, The Netherlands: Reidel.

Veenhoven, R. (2004). *World database of happiness. Continuous register of research subjective appreciation of life* [On-line]. Available at www2.eur.nl/fsw/research/veenhoven/Pub2000s/2004f-full.pdf

Verri, A, Cummins, R. A., Petito, F., Vallero, E., Monteath, S., Gerosa, E., et al. (1999). An Italian-Australian comparison of quality of life among people with intellectual disability living in the community. *Journal of Intellectual Disability Research, 43*, 513–522.

Werngren-Elgstrom, M., Dehlin, O., & Iwarsson, S. (2003). Aspects of quality of life in persons with pre-lingual deafness using sign language: subjective wellbeing, ill-health symptoms, depression and insomnia. *Archives of Gerontology and Geriatrics, 37*, 13–24.

Chapter 6
Too Late to Matter? Preventing the Birth of Infants at Risk for Adult-Onset Disease or Disability

Laura M. Purdy

There are many different views about our duties (or lack of them) toward future people. Some appear to reject altogether the notion that we should to try to influence the future; others think that doing so may be admirable, but not required.[1] But most benefit from and therefore support advances in public health and medicine, such as vaccines and clean water, as well as eyeglasses and prevention or alleviation of an increasingly wide range of health problems. As part of this concern for human welfare, many also believe that children deserve treatment that gives them a good start in life, such as good prenatal care and healthy food.

This consensus falls apart when the only way to prevent disease or disability is preventing certain births altogether. One position is that there is a huge morally relevant difference between preventing disease or disability by treating an individual and doing so by preventing the birth (by any means) altogether. A second position is that it is permissible to try to head off such births, but that the means by which one does so matter. Thus there is a morally relevant difference between preventing certain conceptions (OK) and ending the lives of embryos or fetuses (not OK). A third position is that both preventing conception and ending such lives can be morally justifiable.

My focus here is on the case for preventing the birth—either by avoiding conception or by prenatal screening followed by abortion—of fetuses at risk for serious adult-onset health conditions.[2] I contend that some widely-accepted arguments against these practices are inadequate and that preventing some such births is consistent with compelling and widely-accepted moral principles.

One of the problems with the current state of the discussion is that much of its vitality has been created by vocal disability rights activists who equate preventing births with discrimination against persons with disabilities. However, they frame the debate in terms of "disability," which for them also includes congenital diseases. But in both disability and congenital disease, there is a wide range of possible consequences. Some hardly interfere with daily functioning even with no social support,

L.M. Purdy (✉)
Department of Philosophy, Wells College, Aurora, New York, USA
e-mail: lpurdy@wells.edu

D.C. Ralston, J. Ho (eds.), *Philosophical Reflections on Disability*, Philosophy and
Medicine 104, DOI 10.1007/978-90-481-2477-0_6,
© Springer Science+Business Media B.V. 2010

whereas some create devastating impairments or pain, even with maximal social support. But there is a big difference between being born with (or acquiring) some impairment that is not progressive, that is not painful, and that does not lead to death, and being born with (or developing) a disease with such characteristics. In my view, it is extremely unfortunate that such morally relevant distinctions are tending to get lost.[3] However, this approach is so thoroughly embedded in the literature, as we will see, that it is almost impossible to alter.[4] In fact, at least two diseases have been swept into the discourse (breast cancer (BC) and Huntington's Disease (HD)), inappropriately in my view. In this chapter I will attempt to show why I think this is true.[5]

6.1 What is Conception and Does It Matter?

Many different interventions can prevent a particular birth. Potential parents whose goal is to avoid serious disease or disability can act before conception by refusing to procreate, adopting, or by using donor gametes. After fertilization (but before implantation), it is now possible to use preimplantation genetic diagnosis (PGD) to examine a blastocyst's genotype, discarding those that test positive for a particular disease. Also, once pregnancy has been established, prenatal tests (such as sonograms, chorionic villi testing, and amniocentesis) can be used to attempt to determine whether a fetus is at risk for specific conditions.

Some objections to preventing the birth of children with late-onset problems turn on the moral status of embryos or fetuses. Many think that if they have a full right to life, then killing them is murder, and murder is always wrong.[6] So it is understandable that those who equate abortion with murder would reject any prenatal testing leading to abortion even where it is intended to reduce suffering.[7]

Objections to this position may take several forms. Perhaps the most common "liberal" position is that embryos have at best only very weak moral standing but that fetuses have increasing moral status as they develop, especially after viability.[8] Prenatal testing can generally take place before viability.

In principle, both the position that embryos and fetuses have a full right to life, and liberal objections to it, are compatible with the view that certain conceptions should be prevented. In addition, some objections do not turn on the status of embryos and fetuses. For example, the disability rights view focuses on the claim that preventing these births constitutes unjustifiable discrimination or insult against the disabled. Not all persons with disabilities share this position: some (and their abled supporters) believe that discrimination against them is immoral, and that society must make much greater efforts to provide the conditions in which they can flourish, but that it is not wrong to try to prevent or eradicate disability. I share this position.[9] Concerns about eugenics, effects on the concept of parenthood, and burdens on women also transcend embryonic or fetal status,[10] as does the Parfit/Robertson view of identity[11] that lurks everywhere here, confusing and distorting the overall picture. All of these reflect rather general worries about any attempt to select who gets born and will not be considered here.

Many who accept genetic testing for devastating, early-onset diseases like Tay-Sachs nonetheless reject testing for late-onset diseases. The commonest justifications for this distinction are first, that penetrance of the gene is incomplete, that is, individuals may not develop the disease even though their risk of doing so is well above average. Second, even in the worst case, they will have years of disease-free life. Third, there is always hope that effective new treatments will be discovered. Also, in some cases a disease (or its symptoms) can be diminished or prevented by manipulating the environment. For example, the mental retardation caused by PKU can be prevented by adhering to a stringent diet from birth. These factors are alleged to undermine any claim that individuals might be better off not being born.[12]

However, these considerations do not tell the whole story: I believe that in some cases they are superficial and that further scrutiny would reveal a much less optimistic picture.[13] Moreover, most of their appeal arises from mistaken views of personhood, and a narrow conception of harm that wrongly puts an enormous burden of proof on any attempt to prevent suffering and harm.[14]

6.2 Objections to Selecting Against Individuals at Risk for Adult-Onset Disease: BC and HD

BC and HD differ in some ways but are alike in others. In BC, penetrance is incomplete, and there is a wide range of risk; in HD, penetrance is 100%, and its presence means that individuals will certainly develop the disease if they live long enough. For BC, there are some prophylactic measures that might reduce its seriousness, lower the probability of developing it, or delay its onset; there are as yet none for HD. For BC but not HD, there are treatments that may cure the disease or at least prolong life.[15] Both are awful diseases that cause devastating suffering.[16]

Couples at risk for having children with BC or HD tend to know about them from familial history, and genetic counseling can determine their probability of producing a child with BC or HD. Also, if conception has taken place, the embryo or fetus can be tested for the presence of the relevant genes. In both cases, the only preconception strategy for preventing the birth of a child with the deleterious allele is to break the genetic link between at least one prospective parent and any resulting child by using donor gametes or adopting. After conception, PGD or prenatal testing will reveal whether the embryo or fetus carries the relevant gene. In PGD, the affected embryo is discarded; after ordinary prenatal testing, the fetus is aborted. Failing to implant or discarding embryos is not technically an abortion because medicine defines implantation as the beginning of pregnancy.[17]

6.2.1 BC

In Bonnie Steinbock's view, BC is not a case where intervention on behalf of the possible child is justified for the kinds of reasons stated above.[18] Dr. David King,[19] a disability rights activist, also rejects intervention for late onset diseases, including BC. King is a proponent of the Social Model of disability that locates the suffering of the disabled primarily in the failure of society to adequately meet their needs. This model contrasts with the Medical Model, where suffering is claimed to be caused by disability.[20] It should be noted that this disability rights position is espoused by some percentage of disability rights activists, and it is unknown to what extent it represents others with disability or disease.

According to King, the British Human Fertilisation and Embryology Authority (HFEA) that makes rules about specific reproductive technologies uses "serious medical condition" as its criterion for judging whether it is permissible to test for particular conditions, where such testing may lead to abortion. But King maintains that

> the word 'serious' is not unambiguously defined in medical discourse, and has everyday meanings which interfere. An example is genetic predisposition to breast cancer, which is, in everyday parlance a serious disease, yet it is late onset, the penetrance is still unclear and treatments exist. The word 'serious' is used to offer the public reassurance that PGD is not being used to prevent minor conditions, but in fact it does not offer the necessary protection [from the slippery slope to "designer" babies].[21]

Are Steinbock and King right in believing that it is unreasonable to select against BC? The situation is actually more complicated than is usually recognized. Most cases (90–95%) of BC appear in women with no known risk factors.[22] The rest occur in women who have inherited certain mutations of the BRCA1 or BRCA2 genes. Even then, penetrance ranges widely, from about 35 to 85%.

Part of the explanation for this wide range may be the following. The first genetic studies were done on families where many members developed breast cancer (and the closely-related ovarian and colon cancers); the medical community concluded that having those alleles meant almost certain disease. Later on, more individuals from less thoroughly affected families were tested, and their risk turned out to be significantly lower.[23]

So is it reasonable to intervene here? For those who require certainty before acting, obviously not. But why demand certainty? An 85% risk means that 5 out of 6 women will get the disease, many quite early in life, possibly as early as their late teens or early twenties.

If prospective parents are concerned about their future child's welfare and adopt or use a donor egg or sperm,[24] no harm—and considerable good—is done: an existing child gets a good home or a child is brought to life that is at no special risk for BC. Of course, there are some costs here. Adoption requires resources, and using third-party gametes brings another woman or man into the family, however indirectly. Both wholly or partially cut the genetic link between generations.

PGD and prenatal testing followed by abortion allow the couple to maintain the genetic link, but have their costs as well, including the physical and emotional risks and discomforts borne by the woman.[25] Where penetrance is incomplete, there is also the fear of aborting an individual who would not go on to develop the disease. But as I have argued elsewhere, there is no duty to bring possible people to life.[26] Some will say that that's all very well where the risk is 85%, but what about where it is 35%? Does the difference in statistics change the moral situation? It is true that the costs to prospective parents remain the same, while the potential benefit shrinks substantially. The benefit may still be significant, however, as some reminders about BC will show.

Very little is known about BC prevention.[27] Both nutrition and exercise studies have produced conflicting results. One recent study suggests that lower-fat diets might help, and some studies suggest that exercise might reduce both incidence and recurrence rates. Certain drug regimens (tamoxifen or raloxifene) may also lower risk or recurrence in the short term, but their long-term benefits and risks are unknown. Careful surveillance by mammography, MRI, and clinical examination help to detect cancers while they are still small, but plenty of women have suffered early and deadly recurrences after small but aggressive cancers.

The most drastic measures seem to offer the best hope of prevention: removal of breasts and ovaries. Yet some women who submit to this mutilating approach still develop cancer, for it is currently impossible to remove all breast and ovarian tissue.[28] Moreover, these operations can be emotionally and physically traumatic. Women's breasts are considered a crucial part of sexual attractiveness in many societies, including our own. The loss of ovaries launches women into menopause, which may cause not only further loss of attractiveness, but puts women at higher risk of developing serious health problems such as heart disease or osteoporosis, not to mention loss of childbearing capacity and impaired sexual functioning. In short, these approaches can very significantly lower the quality of life, even in the absence of certainty of benefit.

Many women cannot avail themselves of even these limited and imperfect measures because they are uninformed or because they lack resources. As a result they may feel powerless and anxious. Informed women with resources must decide which measures to undertake, all the while being reminded of the risk they face, a state of affairs especially distressing to those who have watched many family members struggle with BC.[29]

And if a woman does develop BC? Sure, it is "treatable." For those without access to good health care, any but the most slow growing form of the disease will be a death sentence.[30] Even with good health care, "treatable" means just that: it can be treated, but not necessarily cured. The lucky ones get high-quality treatment that is not too painful, are not left with permanent damage, do not experience repeated cancer scares, and go on to live many years without a recurrence, dying of some other problem. For many others, however, the experience is much different. Some drug regimens cause great suffering, and do leave permanent damage (both mental and physical); some women also suffer early refractory recurrences that kill them quite quickly.

The much touted lumpectomy is an improvement over mastectomy for some women, but requires lengthy and sometimes debilitating radiation treatment; reconstruction may help some women's self-image, but is invasive, expensive, and may make it more difficult for local recurrences or new cancers to be detected. As well, discrimination against women with BC (and other serious diseases) is not unknown, causing them to lose their jobs and perhaps difficulty in maintaining or getting health or life insurance.[31]

I believe that anyone who has watched friends or relatives deal with BC would concur with my rejection of King's judgment that BC is not "serious." In any case, I hope that the foregoing will encourage others to become better informed about the horrors of BC.[32] By comparison, overcoming the obstacles to preventing the birth of those at such risk seems trivial.[33]

Unfortunately, decision-making about risk for BC is complicated by the fact that we are still far from having a full understanding of its causes.[34] In fact, given recent developments, I'm not altogether sure how much emphasis should be placed on preventing births, as opposed to studying and controlling environmental factors. But that is a result of lack of knowledge about what policy would be most effective, not any doubts about whether we have a moral duty to try to eradicate the disease.

6.2.2 HD

HD is similar to BC in that it is late-onset, but because it is autosomal-dominant, individuals who carry the gene are certain to develop the disease if they live long enough. It is also different from BC in that there are at present neither preventive measures nor treatments available.

As I have argued elsewhere, HD is dreadful:

> Symptoms and signs develop insidiously, starting at about age 35 to 50. Dementia or psychiatric disturbances (e.g., depression, apathy, irritability, anhedonia, antisocial behavior, full-blown bipolar or schizophreniform disorder) develop before or simultaneously with the movement disorder. Abnormal movements appear; they include flicking of the extremities, a lilting gait, inability to sustain a motor act such as tongue protrusion (motor impersistence), facial grimacing, ataxia, and dystonia.
>
> The disorder progresses, making walking impossible, swallowing difficult, and dementia severe.[35]

The age of onset varies tremendously. It generally ranges from thirty to fifty, but can start as early as the teens, or so late as to be relatively inconsequential. It runs its course in about fifteen years, terminating in death.

Before first the genetic marker, and then the allele itself, were discovered, people in affected families could not know whether they would get HD. If they carried the allele, they would themselves develop it, and each of their children had a 50% risk of getting it too. If they did not carry the gene, they would not develop the disease and their children would be free of risk. In the past, because of the late onset, many would have had children without knowing their own status. Now, they can get themselves tested. A negative result means nothing further is needed to protect their children: a positive result means that their children are at risk. Now we can even determine whether onset is likely to be early or late.

A positive result would obviously be a shock (even if a negative one would be a relief, although it may also bring "survivor's guilt"), and many people do not want to face testing.[36] Such individuals will also be loath to avail themselves of PGD or prenatal testing, as a positive result means that they too carry the allele. However, such individuals could (and should, I would argue[37]) still use pre-conception measures to protect their children.

Some prospective parents are also reluctant to proceed with prenatal testing. Steinbock cites a recent Canadian study where only a small percentage did so. Why? They hoped for a cure in time for their children (Steinbock, 2007, p. 40). King, too, emphasizes the possibility of cures as a reason for not attempting to prevent the births of children at risk for diseases that will cause suffering. He comments:

> Disabled people often complain that genetic research provides tools to prevent disabled people being born, but offers them nothing by way of treatment. Geneticists respond that their aim is not to eliminate disabled people, and that they are confident that research will provide cures, although they acknowledge that this will take time. Very well, let us take them at their word. According to this argument PGD may legitimately be offered for severe early-onset conditions, but for conditions that will not be manifest for several decades, we should certainly not offer PGD, since we must expect treatments to be developed in that time. To offer PGD in such cases would appear to vindicate disabled people's suspicion that these technologies are being developed to prevent disabled people being born (King, 2007).

King's conclusion (that the true goal of reproductive technologies is to prevent disabled people being born because of bias against them) hardly follows from his premise (that geneticists are confident that research will provide cures), in part for the reason he acknowledges: "this will take time." But such an approach makes HD carriers hostage to uncertain medical progress. Nobody can forecast when—if ever—effective treatments or cures will be discovered. Given what is at stake here, in the absence of knowledge that a breakthrough is immanent, feebly hoping for a cure is both morally questionable and strangely relaxed about the welfare and happiness of one's offspring.[38] Suppose that parents took this approach to other elements of safeguarding their children's well-being, such as failing to feed them properly, put them in baby car seats, or take steps to ensure their education? Surely, if and when safe and effective postnatal treatments become available, that will be the time for choosing to go forward with a child at risk.[39]

People also tend to regard the views of at-risk but unaffected individuals as the moral trump here.[40] For example, Steinbock writes: "In 1978, Laura Purdy argued that someone who knew that he or she was at risk for HD had a moral obligation not to reproduce. Had Arlo Guthrie taken her advice, he would not have had his three children. As it turned out, he had not inherited the HD gene from his father, Woody Guthrie. His gamble paid off" (Steinbock, 2007, p. 41).

Given that experiencing something yourself often gives you a new appreciation for what others have lived through, I would be more impressed by hearing from an Arlo that had lived in a different possible world, ten years into HD. That Arlo might have quite a different perspective, and perhaps also regret for having unnecessarily put his children at risk, when he could still have raised children not genetically related to him. I also wonder whether any of those "other-world" children at risk of developing HD, watching their father struggle with the disease and all too aware that his fate might well be theirs, might also have regarded the trade-off—losing the genetic link to their father—as worthwhile. Despite the high value that many place on genetic connections, is that really the most important part of family relationships?

Once again, it would be helpful to try to put oneself in these shoes, thinking realistically (and empathetically) about the questions and issues that people in them face. For those who don't know whether they are at risk, should they devote their early years to preparation for a demanding career, only to find symptoms developing just as they are poised to enjoy its fruits? Imagine a young woman or man, wondering whether to start a family, or noticing the first symptoms just after having done so, knowing that they will not see their children grow up, that they will be burdened with their own increasingly demanding care, and that they will have their own worries about developing the disease.. Imagine, too, the questions facing those who do know. Should they bother to finish school? Should they have children? How can they make sure that they will be cared for once they start developing disabling symptoms? And who will care for family members where several individuals are in various stages of the disease? These questions loom especially large in the U.S., where long-term care is provided either by family members (often dooming their own life-plans, decimating any financial savings), or in poorly regulated and expensive nursing homes paid for either by family members or Medicaid.

6.3 Phantom Children and the Confusions They Sow

This standard wisdom ("hope for the best, and if the best happens, that justifies the risks that were taken") discourages us from incorporating the foregoing kind of thinking into decision-making about reproduction. Prospective parents are considered to do nothing wrong by going forward with children in such cases, and to do something good by giving them a chance at life.

This position—that one is doing children at risk a good turn unless their life can be expected to be so dreadful that they would have preferred not to be born—still colors the whole debate, lending plausibility even to arguments that are logically unrelated to it, such as those mounted by some disability rights activists.

I have argued (and still believe) that this position is based on erroneous premises that obscure their inhumane consequences.[41] Bringing one individual rather than another to life leaves no one hard done by. Those not brought into existence really don't exist anywhere—they are phantoms. "They" are not patiently waiting out in the ether, their claim to existence justified by some misplaced application of "first-come, first-served." Their spokespersons' conviction that "they" therefore have an inalienable right to be born (and that they are discriminated against if they are not) thus depends on metaphysical assumptions that cry out for further scrutiny. In the meantime, it is obvious that, other things being equal, people free of such risk will be happier than those afflicted with it—especially once they realize what has been done to protect them. Do they have less of a right to be here than those not born? No—no more, no less. Rights are not at issue here.

In short, the real question here is not whether people at risk for serious disease or disability wish they'd never been born, as adherents of this view require before attempting to prevent their birth.[42] The real question is whether they would prefer a future free of HD, BC, or some other miserable condition.

Both medicine and public health are predicated on the premise that being free of disease, pain, and disability are highly desirable. In fact, most of us tend to take these things for granted until something goes seriously wrong, but when they do, we usually long for what we have lost, and the loss can have extremely serious consequences for our quality of life.[43]

Do we, as individuals and as a society, really want to abandon the commitment to promoting good health and lack of disability? Do we understand the full implications of rejecting it? For it means abandoning health care, biomedical research, and programs to prevent disease and disability. Vaccinations, dental care, antibiotics, cancer care and far more would simply cease to be priorities. Nor would there be any justification for stringent anti-pollution measures or occupational safety programs. If these values were longstanding, smallpox, polio, and dozens of other diseases would be widespread, and surgeries and drugs to treat and alleviate health problems that we now take for granted would never have been developed. Life wouldn't necessarily be solitary, but it would be poorer, nastier, more brutish, and shorter.

Presumably proponents of the standard wisdom would argue that these implications do not follow. That is, that there is a morally relevant difference between arguing that the risk of serious disease or disability shouldn't be cause for failing to bring the person at risk to life, and arguing that this position is based on the view that there is nothing intrinsically wrong with disease or disability. But the latter is precisely the view articulated by some disability rights activists like King and others.[44] As we have seen, as an adherent of the Social Model of disability, he attributes to society the suffering experienced by people with disabilities. But if there is really is nothing intrinsically bad about having a disability, then any failure of support on society's part cannot be a wrong.[45] So this position is simply incoherent. Disdain for prevention, treatment, or cure of disease or disability does follow from it.

Why do so many, especially progressives, fail to see these points? I believe that the problem is the confusion created by the failure to separate the different arguments here. In particular, the discrimination alleged to be inherent in the attempt to prevent the birth of phantom children at risk for serious disease or disability has somehow become attached to the judgment that it is better to be healthy and able-bodied than to have serious disease or disability. But once the position that it is discriminatory and wrong to prevent the birth of at risk individuals has been laid to rest, either by the kinds of arguments I have mounted or by other approaches,[46] the other arguments need to stand on their own. Other things being equal, realistically— looking at what such risk can do to people's lives (and the social and economic ripples it creates)—it seems awfully hard to make a compelling case that there is no moral difference between being at risk for BC or HD or not.

There are two remaining strands of argumentation that need to be followed up here, however briefly. The first is that even if phantom children are not hard done by as a result of not being brought to life, attempts to eradicate disease and disability are insulting to individuals now living with them, or are perhaps even a form of genocide, diminishing the relevant communities.[47] The second is the reach of the judgment that there is nothing intrinsically undesirable about serious disease or disability.

The objection from insult is unpersuasive, and will not be pursued further here.[48] It does not take much digging to show that the second is equally untenable.

Consider the following thought experiments. Suppose a good (safe, effective, inexpensive) vaccine were found that prevented the expression of HD in children. Would it be wrong to vaccinate them? On the face of it, no, for it addresses the allegedly central distinction between eliminating a disease and eliminating those at risk for it. In fact, it would seem to be just one new weapon against disease. Yet, searching for a vaccine would be just as much a product of negativity about HD as is the attempt to prevent the birth of those at risk for it, and also diminishes the size of the HD community.

If this negative judgment about a condition and the threat to the community of those who have it are not relevant to the vaccine, why not? Is it because those vaccinated still carry their alleles, and so still are, in some sense, part of the HD community, the "honorary" potentially diseased? Would that be because they are still capable of passing the disease on to their children—even if the children also get the vaccination, preventing any further manifestation of the disease? Would it make any difference if the vaccine were inherited, along with the deleterious alleles, eradicating the disease (but not the alleles) even if the children didn't get vaccinated? Or, suppose a still more powerful vaccine were invented, preventing both any expression of the disease and rearranging the alleles into their non-disease producing form, effectively wiping out both disease and community?[49]

To be consistent, disability rights activists would have to be as negative about these vaccines as they are about the attempt to prevent births. But this would show that the central objection of the disability rights position is not to the means used (preventing births) to achieve the goal, but to the goal of preventing disease itself (no matter how it is achieved).

This point also clearly demonstrates how very radical a challenge the disability rights position is to the goals of medicine and public health. It deprives those now suffering from them of any but social measures to improve their well-being, with no emphasis on medical prevention, alleviation, or cure. It must be neutral or even negative about the measures we now enjoy, and must reject attempts to develop new ones.

These conclusions should give pause to liberal and progressive communities whose embrace of the disability rights position has been prompted by solidarity with the language of discrimination, bias, and diversity. Disability rights activists deserve massive support in their battle to eradicate bias and discrimination against those with disease and disability. They also deserve such support in their quest for conditions that permit those with disease and disability to flourish as best they can. But the premises that lead to those conclusions do not support the extension of their argument to the position that it is morally wrong to try to prevent the birth of those at risk for adult onset disease or disability. Moreover, the points they draw on to shore up that unsuccessful extension do have the hitherto apparently unnoticed implication that attempting to prevent disability and disease generally is morally wrong.[50] To say this is to accept and promote vast misery. That alone would be sufficient for rejecting the argument.

Notes

1. These positions are probably generally tied either to ignorance about our power to do so, or to libertarian economic views.
2. This does not mean that most of society's emphasis should be on such genetic issues; to the contrary, I believe that the bulk of action to protect the health and welfare of existing and future individuals should be on political action to end war and poverty, clean up the environment, ensure safe workplaces, encourage healthy lifestyles, and ensure access to high quality health care.
3. For a good discussion that does not do this, see Wendell (2001).
4. I will generally use the now-generic term "disability" myself here, with the understanding that it includes disease, but that it is inappropriate to lump all disease and disability together morally.
5. I have dealt extensively with the morality of preventing disease/disability by preventing some births and will refer to that writing where relevant rather than repeating all the arguments here. See Purdy (1978, 1996c, 1996d). See also a review of Erik Parens and Adrienne Asch, *Prenatal Testing and Disability Rights* (Washington, DC: Georgetown University Press, 2000) in Purdy (2001).
6. For a different perspective, see Thomson (1973), Purdy (1996a, 1996b).
7. Although I believe that this position is weaker than the alternatives, I respect those who hold it consistently. See my Purdy (2005).
8. My own position continues to be more the more radical theory that personhood is not achieved until sometime after birth, although there are other considerations that might well constrain late-term abortions and most infanticide (see Purdy, 1996d). However, I will take the "standard" position as the norm here, and my paper is not directly addressed to those who believe conception marks the beginning of personhood except insofar as they still have a duty to try to prevent certain conceptions.
9. Some, like Asch, distinguish between preventing conception and abortion, but as I have argued elsewhere, this position isn't particularly easy for liberals to support (1996d, esp. pp. 54–55; also my review of Parens and Asch, in Purdy, 2001).
10. See, for example, Inmaculada de Melo-Martin (2006).
11. See Parfit (1984) and Robertson (1994).
12. Steinbock (2007). There is no good way to discuss this issue ("better off not being born"), as all our language seems to imply the prior existence of an individual.
13. This is my focus here, too, although there is also much to be said about these other perspectives. (See Purdy (1996d)).
14. See Purdy (1996d) for a more detailed treatment of what I think are the mistakes being made here.
15. For information about current treatment of and ongoing research on HD, see NINDS (2008). Some symptoms can be alleviated with drugs, but they have serious side effects; there is currently no way to stop or reverse the disease.
16. See, for example, Rimer (2002). This article, and Harmon (2007) (see note 36, below) provide an inside look at the dread inspired by this disease, providing a counterpoint to those who discount or deny it.
17. Pregnancy begins with implantation, according to medicine; "prolifers" hold valuable life begins at fertilization. For some of the many difficulties inherent in such decision-making, see Green (2001, Chap. 2).
18. Steinbock (2007, p. 41).
19. King (2007).
20. The Social Model locates suffering in society, regarding disability as socially constructed and the suffering as caused by society's failure to meet the needs of disabled people. The Medical Model locates the suffering in the condition itself. It seems to me that the best model would combine the two, and that the source of suffering varies in different kinds of cases.

21. King (2007).
22. Of course, men get BC, too, but in much smaller numbers and still less is known about their situation. For the sake of simplicity I focus on women here.
23. For more information about nuances in familial risk, see American Cancer Society (2008).
24. Depending on which family carries the deleterious alleles.
25. In PGD, the risk and discomfort of IVF; in prenatal testing followed by abortion, the risk and discomfort of the tests and of the abortion.
26. See my Purdy (1978).
27. For a brief summary of current information about prevention, see National Cancer Institute (2008).
28. There is ovarian tissue in the abdominal cavity that remains even when ovaries have been removed.
29. It is difficult to overestimate the fear that women in high-risk families live with on a daily basis.
30. Consider the statistics concerning health insurance and morbidity/mortality figures.
31. Even where it is completely irrational. One insurance company would not consider me for long-term care insurance until five years after diagnosis—even though a recurrence would ensure that I would never need long-term care.
32. There are all sorts of resources here, from patient information brochures, films, and first-person accounts. I recommend starting with Batt (1994) and Middlebrook (1996).
33. So trivial, it seems to me, that I wonder about the thought processes involved here. Does our well-documented propensity to discount problems that will occur in the future result in some bias, for example?
34. Other wrinkles here include a recent finding that women in BRCA families who do not inherit the gene still have three times the risk of developing BC compared to women in the general population (Buxton, 2006). Moreover, there is increasing evidence of other genetic contributions to BC, such as the recently noticed PALB2 gene (Buxton, 2007).
35. Merck Manual (2007).
36. See, for example, Harmon (2007). For a brief rundown of the pros and cons of testing for HD, see Drellishak (2007).
37. As I have argued elsewhere, if their fear of the disease is so great that they cannot face knowing their own status, this is all the more reason to try to ensure that their children never are in this situation. See Purdy (1996c, p. 48).
38. See Purdy (1996d).
39. First, any such measures might be burdensome. Even if they are not, circumstances can change so as to make them difficult or unattainable. Suppose that we discovered that HD children can just substitute olive oil for butter to prevent manifestation of the disease. Unless they live where olives grow, it doesn't take much imagination to see that olive oil might become unavailable. Despite the current abundance of olive oil, it wouldn't be at all surprising to see it become unavailable as humanity copes with the twin threats of peak oil and global warming. More demanding treatments or cures might also be so expensive that society cannot or will not provide them.
40. See Purdy (1978), especially with respect to the Joseph Family disease.
41. See Purdy (1996d).
42. I suspect that "yes" may be more often the answer than is acknowledged; however, it is very difficult to get accurate information about such emotional issues.
43. One prominent exception is some persons who are deaf who are committed to the view that deafness is not a disability but rather helps constitute an alternative culture.
44. In part because of these implications, my guess is that only a minority of those with disabilities are represented by disability rights activists such as King.
45. Of course, active social discrimination would still be immoral.
46. See, for example, Brock (1995).
47. Although no one has (to my knowledge) used this word, the appeals to preserve the community suggest its aptness.

48. See Purdy (1996d) and Nelson (2000).
49. Except for future random mutations. In some sense, these thought experiments also erode the apparently clear distinction between preventing the birth of a particular individual and preventing the disease in an existing one.
50. I was groping in this direction in Purdy (1996d), but was not able to articulate it very clearly at that time.

References

American Cancer Society. (2008). *What are the risk factors for breast cancer?* [On-line]. Available at http://www.cancer.org/docroot/CRI/content/CRI_2_4_2X_What_are_the_risk_factors_for_breast_cancer_5.asp

Asch, A. (1989). Can aborting "imperfect" children be immoral? In J. Arras & N. K. Rhoden (Eds.), *Ethical issues in modern medicine* (3rd ed., pp. 317–321). Palo Alto, CA: Mayfield Press.

Batt, S. (1994). *Patient no more: The politics of breast cancer*. Charlottetown, PEI, Canada: Gynergy Books.

Brock, D. (1995, July). The non-identity problem and genetic harms—the case of wrongful handicaps. *Bioethics, 9*(3–4), 268–275.

Buxton, J. (2006). Moderate breast cancer risk in women with negative gene tests. *BioNews, 383* (July 11) [On-line]. Available at http://www.bionews.org.uk/new.lasso?storyid=3243

Buxton, J. (2007). New breast cancer risk gene identified. *BioNews, 390* (August 1) [On-line]. Available at http://www.bionews.org.uk/new.lasso?storyid=3305

De Melo-Martin, I. (2006). Furthering injustices against women: Genetic information, moral obligations, and gender. *Bioethics, 20*(6), 301–307.

Drellishak, R. (2007). *Genetic testing* [On-line]. Available at http://www2.lib.uchicago.edu/~rd13/hd/testing.html

Green, R. (2001). *The human embryo research debates: Bioethics in the vortex of controversy*. Oxford, UK: Oxford University Press.

Harmon, A. (2007). Facing life with a lethal gene. *NY Times* (March 18) [On-line]. Available at http://select.nytimes.com/search/restricted/article?res=FA0B15FC3D540C7B8DDDAA0894DF404482

King, D. (2007). Preimplantation genetic diagnosis and "Slippery Slopes". *BioNews, 407* (May 14) [On-line]. Available at http://www.bionews.org.uk/commentary.lasso?storyid=3441

Merck Manual. (2007). [On-line]. Available at http://www.merck.com/mmpe/sec16/ch221/ch221e.html#sec16-ch221-ch221e-1299

Middlebrook, C. (1996). *Seeing the crab: A memoir of dying*. New York: Basic Books.

National Cancer Institute. (2008). *Breast cancer PDQ: What is prevention?* [On-line]. Available at http://www.cancer.gov/cancertopics/pdq/prevention/breast/Patient

National Institute of Neurological Disorders and Stroke. (2008). *NINDS Huntington's disease information page* [On-line]. Available at http://www.ninds.nih.gov/disorders/huntington/huntington.htm#Is_there_any_treatment

Nelson, J. (2000). The meaning of the act: Reflections on the expressive force of reproductive decision making and policies. In E. Parens & A. Asch (Eds.), *Prenatal testing and disability rights* (pp. 196–213). Washington, DC: Georgetown University Press.

Parfit, D. (1984). *Reasons and persons*. Oxford: Clarendon Press.

Purdy, L. (1978). Genetic diseases: Can having children be immoral? In J. L. Buckley (Ed.), *Genetics now* (pp. 25–39). Washington, DC: University Press of America.

Purdy, L. (1996a). Abortion and the argument from convenience. In L. Purdy (Ed.), *Reproducing persons* (pp. 132–145). Ithaca, NY: Cornell University Press.

Purdy, L. (1996b). Abortion, forced labor, and war. In L. Purdy (Ed.), *Reproducing persons: Issues in feminist bioethics* (pp. 146–160). Ithaca, NY: Cornell University Press.

Purdy, L. (1996c). Genetics and reproductive risk: Can having children be immoral? In L. Purdy (Ed.), *Reproducing persons: Issues in feminist bioethics* (pp. 39–49). Ithaca, NY: Cornell University Press.

Purdy, L. (1996d). Loving future people. In L. Purdy (Ed.), *Reproducing persons: Issues in feminist bioethics* (pp. 50–74). Ithaca, NY: Cornell University Press.

Purdy, L. (2001, October). Review of *prenatal testing and disability rights*, by Erik Parens and Adrienne Asch. *Social Theory and Practice, 27*(4), 681–687.

Purdy, L. (2005, October 29). Sex, lies, and the religious right: "Culture of life" or culture of misery. *The new enlightenment.* Unpublished paper, Amherst, NY.

Rimer, S. (2002). A deadly disease destroys patients and families. *NY Times* (June 24, 2002) [On-line]. Available at http://query.nytimes.com/gst/fullpage.html?res= 9C05E0D61E3FF937A15755C0A9649C8B63&sec=health&spon=&pagewanted=all

Robertson, J. (1994). *Children of choice: Freedom and the new reproductive technologies.* Princeton: Princeton University Press.

Steinbock, B. (2007, December). Prenatal testing for adult-onset conditions: Cui bono? *Reproductive Medicine Online, 15* (Ethics, Bioscience and Life, Supplement 2), 38–42.

Thomson, J. (1973). A defense of abortion. In J. Feinberg (Ed.), *The problem of abortion* (pp. 121–139). Belmont, CA: Wadsworth.

Wendell, S. (2001). Unhealthy disabled. *Hypatia, 16*(4), 17–33.

Chapter 7
To Fail to Enhance is to Disable

Muireann Quigley and John Harris

7.1 Introduction

It is sometimes said that while we have an obligation to cure disease and prevent
or ameliorate disability, we do not have an obligation to enhance or improve upon
a normal healthy life. Indeed it is often said that not only do we have no obligation
to enhance, we have a positive duty not to do any such thing. Enhancements are
often, perhaps most usually considered anathema (Sandel, 2004). In this paper we
argue against this assertion and maintain that not only does such an obligation exist,
but also where it is possible to confer upon one's child certain health advantages or
enhancements, then to fail to enhance is in fact to disable them.

7.2 What is Disability?

A pervading fallacy in the writings on healthcare is to assume that disability, and
disease in general, can only be defined in relation to so-called normalcy, normal species functioning, or species-typical functioning (Daniels, 2000, 1996, 1985
Boorse, 1981). Such definitions rely on two erroneous assumptions. The first is that
we can actually give a meaningful account of what it means to function within the
normal, parameters of our species; and the second is that we can quantify this and
thereby know when we encounter those who fall outside its scope. In relation to
disability it is generally those who then fall below the minimum standard for normalcy that are considered to be disabled. However, disability does not need to,
and ought not to be defined relative to normalcy. The reason for this is twofold.
The first reason is that a person can be "normal," that is, normal in the sense of
being what people would consider to be normal, and still be disabled. To illustrate, consider the following scenario. Suppose due to further depletions to the

M. Quigley (✉)
Institute for Science, Ethics and Innovation, School of Law, University of Manchester,
Manchester, UK
e-mail: Muireann.Quigley@manchester.ac.uk

D.C. Ralston, J. Ho (eds.), *Philosophical Reflections on Disability*, Philosophy and
Medicine 104, DOI 10.1007/978-90-481-2477-0_7,
© Springer Science+Business Media B.V. 2010

ozone layer, all white-skinned people were very vulnerable to skin cancers on even slight exposure to the sun, but brown and black-skinned people were immune. We might then regard whites as suffering substantial disabilities relative to their darker-skinned fellows. And if skin pigmentation could be easily altered, failure to make the alterations would be disabling (Harris, 2001, p. 384). We will return to the issue of enhancements later. For the moment it is sufficient to note that in such circumstances whites might have disabilities relative to blacks even though their functioning was quite species-typical or normal. And of course disability can only be defined relative to some other condition not so defined! The second reason is that the moral imperative, and the most usual moral motive, for medical interventions is not to return an individual to "normal" functioning, but to change their condition where possible for the sake of the harms these changes will prevent or palliate and the goods that this will bring about. When someone is suffering we do not ask "is this suffering a normal part of the human condition?", we ask (or we hope any decent person would ask) "can we do anything about it?" What then is the alternative?

Harris has defined disability or disease as a physical or mental condition that someone has a strong rational preference not to be in and one that is, in some sense, also a harmed condition (Harris 1992, Chap. 4; 1993, p. 80). So it is a condition that you, or I, or anybody else, all other things being equal, would rationally prefer not to have, because having or being in that condition in some way harms us. A harmed condition is the sort of condition which, if an unconscious patient presented with it to the emergency department of a hospital, and the condition could be easily and immediately reversed, the medical staff would be negligent were they not to attempt reversal. A harmed condition is also the type of condition which, if a pregnant mother knew that it affected her fetus—and knew also that she could remove the condition by, say, simple dietary adjustment—then to fail to do so would be to knowingly harm her child.

We can articulate this harm in another way and say that a person is harmed when their interests are thwarted, and where there is a remedy, failure to implement it is a necessary condition of the harm. Where the costs of preventing this frustration of interests are small compared with the harms caused it is difficult to think that there might be an excuse not to prevent the harm being caused.[1] These interests can be understood in two ways. The first is in terms of the physical, where a person has an interest in not experiencing suffering and pain; thus they are harmed if something occurs that causes them to experience suffering and pain. The second is in terms of life's opportunities, where a person has an interest in having open to them the variety of opportunities and options that are open to other people. Thus they are harmed if "important options and experiences" (Harris, 1993) are foreclosed to them. Both of these types of harms are surely ones which we would all have a strong rational preference to avoid if we could.

This account of disability does not give its definition in relation to normalcy or any sort of species-typical functioning. It is defined relative both to one's rational preferences and to the alternatives. When we use this account of disability we can see that the issue is not about a person's being "normal" or "abnormal," but about them being worse off than they otherwise *could have been*. This being true, it is also true that where we can take steps to prevent people being in a harmed state and being worse off than they otherwise could have been, there is a moral imperative to do so.

Let us take the example of a woman who is trying to become pregnant. This mother already has one child affected by spina bifida (a neural tube defect which occurs very early in pregnancy and involves the incomplete development of the brain, spinal cord, and/or their protective coverings). She is aware that evidence shows that taking folic acid pre-conception will reduce the risk of neural tube deficits in any resulting child. She refuses either to take folic acid supplements or to follow a folic acid rich diet. Is she morally responsible for the contribution that this conduct makes to her child being born with spina bifida? The answer must surely be yes.

Of course we all could have been worse off than we are now or, indeed, better off than we are now. A multitude of factors have, or could have, impacted on our current state of functioning. These include pre-, peri-, and post-natal factors: when our parents tried to conceive, when they were successful, what our mothers ate during her pregnancy, what illnesses we have had since we were born. Among the pre-natal variants there is even the possibility that I may never have existed at all. With such an array of variables is it reasonable for parents to be held morally responsible for the condition, harmed or otherwise, of their children? Surely they cannot be held accountable for all harms that accrue, pre-natally and in early childhood, to their children?

It is true that our parents are causally responsible for the condition that we are born in (and any harms that stem from this)—how could it be otherwise? However, they are surely not *morally* responsible unless they were (1) aware that they were likely to transmit those harms (or benefits); and (2) aware that they might have a better alternative child or better possible alternative child. Of course, given some of the variables mentioned above, there could always have been a better *possible* alternative child, so we need a third condition and that is that our parents could *realistically* (at an acceptable balance of cost and benefit) have produced that child instead. This is necessarily a vague notion, but it is one we humans employ all the time. When we ask "could the accident have been realistically prevented?" we are asking not only whether or not it could *actually* have been prevented, but whether or not it would have been reasonable to expect those who could have prevented it to have done so, bearing in mind the costs of so doing. In the case of spina bifida described above we can see that indeed the mother could have *realistically* produced that child.

The argument presented here is not the claim that we ought to prevent harm and disability only where the life produced would be a life that is so bad it would not be worth living. The argument is simply that we have reasons not to start out life, or cause life to be started, with any unnecessary disadvantages and we have reasons to prevent harm, however slight. Given that this is true it is also true that we have good reasons to start out life with any advantages or enhancements that we can, however slight.

7.3 What Are Enhancements?

There are two questions which need to be answered here. The first is what exactly an "enhancement" is and, correlatively, what an enhanced condition is; and the second, whether or not it is actually better to have an enhancement. The short answer to the second of those is that it must be better to have an enhancement; if it were not, then it would not be an enhancement. The first question is more difficult to answer.

When talking about enhancement, we encounter the same fallacious thinking that we did with disability—namely, that enhancement necessarily needs to be defined relative to normalcy. This approach is neither relevant nor appropriate. Like disability, enhancement does not need to be and ought not be defined in this way. We can say that an enhancement is anything that makes a change or a difference for the better. And, like disability, we can define an enhanced condition without recourse to concepts of normalcy or normal species functioning.[2] However, recourse to such a concept is not needed; thus we can again define it relative to a person's rational preferences and to the alternatives. An enhanced condition is a physical or mental condition which one would rationally prefer and is not a harmed condition. This account is about individuals being better off than they otherwise *could have been.*

This definition, therefore, necessarily includes simple everyday medical interventions such as spectacles, contact lenses and laser eye surgery, statins for reducing cholesterol and preventing heart disease, and insulin for diabetes, but will also encompass (eventually) technologies not yet available, such as gene therapy. Indeed it can be seen to include anything which could be deemed to give us more and better health.

One might point out that this definition applies not only to so-called enhanced health states but to so-called normal health conditions, too. This is because our so-called "normal" states of health are also surely conditions that one would rationally prefer to be in. This, however, simply highlights another strand of flawed reasoning when it comes to enhancements. This is the reasoning employed by those who claim that there is a coherent or usable distinction with moral relevance between medical treatments on the one hand, and medical enhancements on the other.

7.4 The Treatment/Enhancement Distinction

It is sometimes said that while we have an obligation to cure disease—to restore normal functioning—we do not have an obligation to enhance or improve upon a normal healthy life, that enhancing function is permissive but could not be regarded as obligatory. Some would even go as far to say that enhancing function is not a morally permissible activity. While this stance is questionable in and of itself, there is an even more fundamental flaw to these assertions. The mistake made is twofold. First, there is the erroneous assumption that there is actually some clear-cut dividing line between that which is a treatment and that which is an enhancement. Second, it is also assumed that this dividing line, were it to exist, acts as some kind of moral partition, and that being assigned to one side or the other is all it takes to affect the morality of the intervention. In order to demonstrate the first mistake, let us look at two examples. The first example shows how in different people the same intervention could be considered to be a treatment in one and an enhancement in another, and asks if this is really coherent. The second example is of a medical intervention which is commonly considered to be a treatment but which, when examined, is really better considered an enhancement.

Our first example is that of patients in Chronic Renal Failure (CRF). Such patients are often given a drug called erythropoetin (Epo). This is a substance whose production is impaired in CRF but which is needed for the production of red blood cells. It is also the same substance that is referred to in athletics in the so-called "blood doping" scandals. Epo is used in that context because it increases the production of red blood cells and consequently increases the oxygen-carrying capacity of the blood and thereby improves the stamina and capacity of the athlete. Notably, training and living at high altitudes also has the same physiological effects. It might be argued that the two uses of Epo described above are distinct, that one treats an ill person while the other enhances a "normal" person and, for that reason, the first is a morally acceptable practice while the second is not.[3] Regardless, it is still true that the person in Chronic Renal Failure who receives Epo has their function enhanced relative to the alternative, the alternative being not receiving it.

Our second example is one of the best examples of where the treatment/enhancement line is blurred. This is the case of immunizations. Most of us have received a plethora of immunizations over our lifetimes, from childhood vaccinations such as polio, tetanus, and Haemophilus influenza B to those, such as the Japanese encephalitis and rabies vaccines, that we get when travelling to far-flung places abroad. Are these treatments? They cannot surely be construed as such. If treatment, in the traditional sense, is that which is administered to cure an already present disease, then immunizations are not treatment. They are given on a prophylactic basis prior to any contact with the particular disease. And how do they work physiologically? Well they work simply by augmenting, or *enhancing*, our natural immune function so that it can better fight the disease if we come into contact with it. Therefore, it would not be true to say that we do not already engage in a variety of enhancement-type practices in medicine, because we clearly do. However, it would be almost unthinkable for those opposed to enhancement to suggest that these immunization programs ought to be either banned, or not considered morally obligatory, on the grounds that they are enhancements.

While there have been various attempts to demonstrate a treatment/enhancement distinction, one of the most notable comes from Norman Daniels. In his paper "Normal Functioning and the Treatment-Enhancement Distinction" (Daniels, 2000) he attempts to answer criticisms such as those mentioned above. He notes that the criticisms against such a distinction are that "it is difficult to draw, that it does not give us the boundary between what is obligatory and non-obligatory in medical interventions, and that it leaves us with hard cases that make the distinction seem arbitrary" (Daniels, 2000, p. 309). We agree that these are the criticisms, but not with Daniels' purported refutation of them. While we cannot go through each of them here, we consider one of his central rebuttals below.

Daniels claims that:

> it is our norms and values that define what counts as disease, not merely biologically based characteristics of persons. . . .. Pointing to the line between treatment and enhancement is not, then, pointing to a biologically drawn line but is an indirect way of pointing to the valuations we make (Daniels, 2000, p. 313).

If this is true, it raises the question as to what the relevant evaluation is. The answer must surely be that we have a strong rational preference not to be harmed in a particular way and it is this that leads us to make the evaluation that draws the line. Occam's razor, that wonderfully enhancing surgical tool, comes to our aid here. We do not need to use our values to create a spurious distinction—the normal/abnormal divide—which Daniels insists is the route to distinguishing therapy from enhancement. Occam's razor enhances both the argument and the relevant distinctions, and leaves us with a process that leads not from our norms, through a definition of disease to a distinction between therapy and enhancement, but one which moves directly from our values to a rejection of harm and an acceptance of benefit, whether called therapy or enhancement. This shows that the enhancement/therapy distinction does not come via conceptions of normalcy at all but from the fact that we value minimizing harm. Normalcy plays no part in the definition of harm and therefore no part in the way the distinction between therapy and enhancement is drawn.

Moreover, if our "norms" and "values" are apt to change, then this line between treatment and enhancement, biological or not, is also apt to change. These changing norms and values are not just social, but are also medical. For example, one hundred years ago the rate of mortality from tuberculosis was almost one hundred percent but now not only can it be treated effectively with a host of antibiotics, we can also be immunized (that is, have our immune systems enhanced) against it. Situations in the past where the medical inevitability was once death and suffering no longer obtain. To base the justification of the enhancement-treatment distinction on a morally arbitrary concept is simply to reinforce the moral arbitrariness of the distinction itself.

Let us now turn to the second line of mistaken reasoning employed by those who oppose enhancements. Supposing that we could adequately demonstrate that there is a clear distinction between treatment and enhancement and between correcting a deficit and enhancement, it is still not evident that such a distinction is actually morally relevant. For such a distinction to hold any sort of moral sway one must establish why enhancements in and of themselves are morally suspect. Leaving aside the fact that by definition enhancements must be good for us (if they were not they would not be enhancements), the ethics of medical interventions ought not to be judged simply on the basis of their membership in a category. Their ethics ought to be judged on the basis of whether or not they prevent or ameliorate harm and suffering, and whether or not they can be said to benefit the individuals who may avail themselves of such interventions. Whether a medical intervention is deemed to be a treatment or an enhancement, it is its relation to this harm or benefit that is the moral crux of the matter.

It often appears that people are objecting to the method of enhancement rather than to the enhancement itself. We do not object to protection from certain diseases through vaccination programs, but objections are raised when it is suggested that the same result could be achieved through gene insertion. If the same medical result can be achieved by two different methods, then why should we object to one and not the other? Is there really a difference between a vaccine that confers immunity to a certain disease and that immunity being conferred by a so-called enhancement technology such as gene therapy? Suppose some pre-implantation embryos could safely be given a genetically engineered condition which conferred immunity to many major diseases—HIV/AIDS, cancer, and heart disease, for example. We would, it seems, have moral reasons to prefer to implant such embryos given the opportunity of choice. There does not appear to be a clear moral distinction between achieving such effects through immunization programs and achieving them via other technologies such as gene therapy.

Thus, there appears neither to be a clear dividing line between what we consider to be treatment and what we deem to be enhancement, nor usually between the methods used to achieve such ends. They represent a series of continua. Where pain, suffering, or harm can be ameliorated or avoided then we are under a moral obligation to do so and this obligation is present whether it means eradicating disability and disease or enhancing "normal" functioning. As such, the burden of proof as to the immorality of enhancement technologies must lie with those who oppose them.

7.5 To Fail to Enhance is to Disable

We have seen that both enhancement and disability can be defined relative to a person's rational preferences and to the alternatives. As such, a disability is a physical or mental condition that someone has a strong rational preference not to be in and one that is, in some sense, also a harmed condition; and, correlatively, an enhanced condition is a physical or mental condition for which one has a rational preference and which is also not a harmed condition. These descriptions do not rely on a person being categorized as "normal," "abnormal," or even "supranormal," but are, instead, about them being worse or better off than they otherwise *could have been*. For that reason they are considered to be in a disabled condition if they are worse off than they could have been, and are in an enhanced condition if they are better off than they could have been. There is also, it seems, a continuum between harms and benefits such that the reasons we have to avoid harming others or creating others who will be harmed are continuous with the reasons we have for conferring benefits on others if we can. In short, to decide to withhold a benefit is in a sense to harm the individual we decline to benefit. For these reasons (how could it be otherwise?), any parent who fails to confer health advantages or enhancements on their child, where they realistically could have done so, has disadvantaged their child and has, therefore, knowingly harmed them.[4]

If we look again at the example of the embryos with a genetic condition that confers immunity to many major diseases, HIV/AIDS, cancer, and heart disease, it is plain that any rational person would prefer to be in this genetic condition (whether naturally occurring or achieved through gene insertion), and it would be difficult to see how being in such a condition could ever be considered to be harmful. The child whose parents did not take steps to furnish them with such health benefits, if and when they become reasonably available, has been harmed. And the parents are not only causally responsible, but also morally responsible for this harm if they allow it to occur. Of course, as mentioned earlier, they must have been (1) aware that they were likely to transmit those harms; (2) aware of the possibility of gene modification (which would of course have to be safe and affordable); (3) aware of a better alternative child or better possible alternative child; and (4) could, *realistically*, have produced that child instead. If the parents knowingly refused to enhance their embryos it is clear that they would in fact have disabled the resulting child.

Likewise if we look again at our spina bifida example we can see that to fail to take folic acid pre-conception, where it is known that this reduces the risk of neural tube defects, is to disable the resultant child who has this disease. It is disabling both relative to the alternatives, and to other individuals. For this reason, if a pregnant mother can take steps to cure or prevent a disability that affects, or might affect, her fetus, she should certainly do so—for to fail to do so is to choose to handicap her child. Correlatively, if a pregnant mother can take steps to enhance or confer health benefits upon her child she should certainly do so; for to fail to do so is to choose to knowingly handicap her child.

To further illustrate our point, we ask the reader to imagine a world which had not been protected against smallpox, polio, and diphtheria, or against measles, mumps, and rubella for that matter. These are all diseases which carry a high rate of morbidity and mortality for those who are unprotected. Now imagine that the methods of immunization had been developed but were never implemented. What would or should we think about those people in our hypothetical world who were not protected against these when they could have been, either through vaccination or other methods? Ask yourself: were those people not disabled?

7.6 Conclusion

In this chapter we have questioned whether there is a morally relevant distinction between medical treatments and enhancements—and, hence, between repairing dysfunction and enhancing function—and have argued that there is not. Claims that one may be a legitimate activity while the other is not were shown to be incoherent. Given that this is true we argue that it is also true that where we can confer upon our children health advantages or benefits, that to fail to do so is to knowingly disable them. On the arguments set out in this paper it is hard to see failing to enhance as anything but disabling, and this is true both by definition and by logical analysis.

Notes

1. For a more detailed account of the ways in which this balance between costs of intervention and costs of failure to intervene can be resolved see Harris (1980).
2. Even if it could be shown that some conception of normalcy was valid here, it surely could never function as anything other than a minimum rather than a maximum standard.
3. Argument regarding "fair play" in sports aside.
4. See Glover (1990, pp. 92–112) for a good discussion of acts and omissions.

References

Boorse, C. (1981). On the distinction between disease and illness. In M. Cohen, T. Nagel, & T. Scanlon (Eds.), *Medicine and moral philosophy* (pp. 11–13). Princeton, NJ: Princeton University Press.

Daniels, N. (1985). *Just health care*. Cambridge: Cambridge University Press.

Daniels, N. (1996). *Justice and justification*. Cambridge: Cambridge University Press.

Daniels, N. (2000). Normal functioning and the treatment-enhancement distinction. *Cambridge Quarterly of Healthcare Ethics, 9*, 309–322.

Glover, J. (1990). *Causing death and saving lives*. London: Penguin Books.

Harris, J. (1980). *Violence and responsibility*. London: Routledge and Kegan Paul.

Harris, J. (1992). *Wonderwoman and superman*. Oxford: Oxford University Press.

Harris, J. (1993). Is gene therapy a form of eugenics? *Bioethics, 7*, 179–189.

Harris, J. (2001). One principle and three fallacies of disability studies. *Journal of Medical Ethics, 27*, 383–387.

Sandel, M. J. (2004, April). The case against perfection: What's wrong with designer children, bionic athletes, and genetic engineering. *Atlantic Monthly, 292*, 50–54.

Chapter 8
Rehabilitating Aristotle: A Virtue Ethics Approach to Disability and Human Flourishing

Garret Merriam

8.1 Introduction

Aristotle would hardly be the first person one would think of when looking for an enlightened understanding of disability. The ancient Greek world in general did not have what we today would consider a progressive attitude towards disability, of course, but it is likely that Aristotle's thought set the standard for conceptualizing disability for more than a millennium to follow. He explicitly endorsed, in the *Politics*,[1] the common Greek practice of leaving "deformed" babies to die of exposure and is reputed to have claimed that those "born deaf become senseless and incapable of reason."[2] Given Aristotle's view of women and "natural slaves," it hardly comes as a shock that he held a less than estimable view of persons with disabilities.

Yet despite this demeaning attitude towards disability, the aretaic ethics that Aristotle pioneered can provide us with considerable resources for making sense of some moral quandaries that complicate thinking about disability. In much the same way that some feminist thinkers have used Aristotle's ethics as a point of departure for developing the Ethics of Care, a rehabilitated Aristotle (with some assistance from the Stoics) can be a starting point for making sense of disability in the modern context.

In this chapter, I will attempt to sketch a modern Aristotelian moral theory with the particular intention of illuminating our understanding of disability and its relationship to human flourishing. Specifically, I will appeal to the Aristotelian notion of practical wisdom and apply it to some paradigm cases to illuminate the relationship between disability and human flourishing. I will conclude by applying the theory thus far developed to make moral assessments of individuals who deliberately choose to create children with disabilities. Throughout I will be focusing on questions of individual human flourishing and how to best understand its relationship to disability; I will not be considering issues of social justice or public policy.

G. Merriam (✉)
Department of Philosophy, University of Southern Indiana, Evansville, Indiana, USA
e-mail: gamerriam@usi.edu

D.C. Ralston, J. Ho (eds.), *Philosophical Reflections on Disability*, Philosophy and Medicine 104, DOI 10.1007/978-90-481-2477-0_8,
© Springer Science+Business Media B.V. 2010

8.2 Rehabilitating Aristotle

8.2.1 Aristotle and Species Essentialism

Aristotle's view of disability was predicated on his understanding of biology, the vast majority of which we now understand to be false. In particular, his conception of species essentialism—the idea that there is an eternal fundamental nature that constitutes the species—was overthrown by the advent of Darwinian evolution. Rather than there being a single core essence of *human being* that sets the gold standard by which all individual humans are to be measured, the species *homo sapiens*, like all other biological species, is dynamic, in flux. Accordingly, there is no single, absolute, metaphysical, archetypal human being by which the rest of us are to be measured. This revelation was a key step in the long process of discarding the understanding of disability as a purely biological concept and shifting towards a perspective that sees disability as socially constructed.[3]

The Darwinian revolution caused some theoretical difficulties for Aristotelian ethics because of the fact that Aristotle's conception of virtue was dependent on his misinformed biology. Without an objective and absolute understanding of what it means to flourish as a human being, how can modern virtue ethicists provide a stable metaethical foundation for virtue? But since virtue ethics was out of fashion with moral philosophers at the time of Darwin this problem failed to connect with a large body of scholars. When virtue ethics came back into vogue in the 1950s the lessons of Darwin had been thoroughly absorbed into the intellectual soil, and new theories of virtue (Aristotelian in spirit if not in letter) that were free of the tethers of species essentialism were duly constructed.

8.2.2 Aristotle, Boorse and the Biostatistical Theory of Disease

Even post-Darwin, species-standardized models of flourishing have had staying power. Consider, for example, the biostatistical theory of disease (BST) presented and defended at length by Christopher Boorse. According to the BST human health is defined as "the absence of disease; [whereas] disease is only statistically species-subnormal biological part-function."[4] The BST has both key differences and important similarities with Aristotle's view of flourishing. First, Boorse *does not* attempt to give a general theory of human flourishing the way Aristotle does (and deliberately avoids value-laden concepts like "living well").[5] At the same time, however, Boorse is trying to give a general theory of human health, which is for Aristotle[6] an important component of flourishing.[7] Another important difference is that Boorse does not fall prey to Aristotle's fallacy of species essentialism (something any respectable thinker since Darwin must avoid). This difference notwithstanding, the biostatistical theory does claim that health and disease are to be defined in terms of how a given subject measures up to the "average person." It is this last similarity that is the most important for our purposes here; both Aristotle's theory of flourishing and Boorse's theory of health use a species standard as their common metric. As a result neither can properly distinguish between disease and disability, between conditions that inhibit flourishing and ones that imply different criteria for flourishing.

There is a form of unjustified discrimination going on in both the literal Aristotelian account of flourishing, and the Boorseian account of health. Whether or not a given individual is physiologically, perceptually or cognitively similar to the species-norm is an arbitrary standard for living well. We can perhaps see this arbitrariness more clearly if we recall a past example of such a spurious standard, namely Aristotle's notion that women are inferior because they do not live up to the norm set by men.[8] We could choose to define the term "flourishing" in exclusively male terms, but is there any independent reason to do so? Would it help us garner any profound insights into the nature of the good life? Is there any theoretical advantage to using men as the standard, as opposed to, say, women, or transsexuals for that matter? In retrospect, we see Aristotle's decision to use the male as the benchmark for human flourishing not as a moral or scientific insight, but rather as an expression of his own bias and chauvinism.

Something similar is happening in any theory that, akin to Boorse's, takes biostatistical normalcy as a prerequisite for human flourishing. To simply *define* flourishing in terms of the "normal person's" capacities displays a chauvinism all its own. The lingering ghost of Aristotelian essentialism still haunts us, convincing us without rational justification that the species-line is, per se, the relevant one. Are there any non-arbitrary reasons for using the species-average (i.e., the species-typical statistical norm) as the basis for our conception of flourishing, as opposed to, say, the genus, the phylum, or even the kingdom? It is certainly the case that, in general, we have more in common with other members of our own species than we do with members of other species. But then again, in general, we have more in common with members of our own gender than we do with members of another gender. Why not abandon the species-average and declare that the barometer for our flourishing is to be calibrated to a given gender-average? There is certainly a relationship between a person's physiological and cognitive condition and her flourishing (just as there is a relationship between a person's gender and her flourishing) but to use the species-as-a-whole as the yardstick to measure that relationship is just as arbitrary as using men as the yardstick to measure women.

8.2.3 A Better Understanding of Biology and Flourishing

A better reconstruction of the relationship between biology and flourishing would assert that when making assessments of how well one lives, a new question needs to be asked. Instead of wondering "how does this individual compare to a species-norm in terms of the capacities necessary for flourishing?" we must instead ask, *given the individual circumstances of this person's life, are they living well, or living poorly?*[9] This question is decidedly closer to the Stoic understanding of virtue, which views flourishing primarily as a matter of what you do with the circumstances of your life, than it is to the Aristotelian view, which places much more weight on external circumstances. To rehabilitate Aristotle involves not only divesting him of his mistaken biology, but also giving him a dose of Stoic fortitude. These two points go hand in hand; once we have removed the species-standard as a barometer for judging human flourishing we are left with nothing other than individual circumstances. When assessing those circumstances we must take into account the biological facts of the individual, as well as the cognitive, psychological, social and esoteric factors that come together to compose their life. This vision of *eudemonia* still adheres to the Aristotelian notion that "anything that lives can live well or live poorly" while also avoiding the species essentialism that plagues Aristotle's literal theory.[10]

It should be explicitly noted that heretofore I have been using the term "disability" in the broad sense that conflates the physiological considerations with the social considerations that prevent full participation in society. I have done this because it would be anachronistic to make such a distinction when speaking of Aristotle, as the division is a contemporary one. In recent years disability rights advocates have labored to distinguish these two concepts, using the term "impairment" to refer to the biomedical considerations and "disability" to refer to the social ones. But despite the considerable emphasis placed on this distinction by disability rights activists, I wish to contend that *at least in the context of individual flourishing*, concepts like "disability" and "impairment" are not terribly useful. The basic concepts are perhaps valuable for the sake of conceptual clarity, but they don't help us to answer any of the perplexing moral questions we're faced with in the context of these issues. Hence, the debate over whether or not disability is a biological condition (as Aristotle and Boorse both seem to contend) or whether it is a social construction (as most disability rights activists contend) is, in this limited context, more or less beside the point.[11] While both biological and social dimensions will be relevant, neither will be dispositive, nor even generally indicative of *eudemonia*. Whether or not one is disabled[12] (however we choose to construe that concept) will not, by itself, tell us much about whether or not that individual is flourishing. One can be disabled on either account and yet still flourish, just as one can be non-disabled and still fail to flourish.

8.2.4 Practical Wisdom, Judgment and Disability

It's not hard to see how this understanding of human flourishing would apply to the issues that typically fall under the umbrella of "disability." *Whether or not an individual has a given set of abilities, is not the key question; what matters is whether or not, given the particulars of their circumstances, they are living well*, where "living well" is not determined by a simple species-standard, but rather based on a considered application of *phronesis* ("practical wisdom") to those circumstances. The use of this practical wisdom takes the place of the species-standard metric employed by the theories of Aristotle and Boorse. To know if someone is living well we do not need to know how they measure up to others of their kind, we merely need to identify and understand the relevant factors in that individual's life and judge them appropriately.

The role of *phronesis* in virtue ethics is well established,[13] and it plays a central role in the application of virtue ethics to practical moral problems. In contrast to some other moral theories, virtue ethics has no single supreme moral principle, nor is there a "moral calculus" with a given set of variables into which moral values are plugged, in turn yielding a definite, clear-cut solution to every moral conundrum. Virtue ethics tends to see the moral world as too complex, too "fuzzy" to submit to such formulaic approaches. Rather than relying on abstract standards that apply in all circumstances, virtue ethics maintains that making moral choices by necessity involves making perspicacious judgments.[14]

The idea of elevating and empowering judgments about the lives of persons with disabilities might make some nervous. Historically, most such judgments passed by people in positions of authority have at best marginalized persons with disabilities and at worst deemed them "life unworthy of life." The history of the eugenics movement is a catalog of judgment after judgment condemning the impaired as subhuman. And as I have already noted the primary apologist for judgment, Aristotle himself, came to rather disgraceful conclusions regarding disability. Given the incredibly poor track record that judgment has in this respect, should we not take a rather cynical attitude towards the idea that it can be relied upon to parse the issues at hand? This skepticism may drive some towards a more absolutist approach to these issues, one that does not rely so much on the judgment of individual agents (such as that of rights theory, for example.)

No one with an awareness of the dark history on this subject can dismiss this objection out of hand. Nonetheless, to distrust the faculty of judgment on the basis that it has been misused in the past is not only specious, but also self-contradictory. When we reflect on the prejudice of others towards persons with disabilities, it is precisely our own judgment that we employ when evaluating theirs. Moreover, the history in question more often than not involves blanket condemnation of whole classes of people, rather than a nuanced, particularistic assessment of the merits of individual cases, which I am advocating here. The problem is not judgment itself, but rather judgments made in ignorance, judgments tainted by bias or ideology, judgments made by those with a noted lack of moral wisdom. A person who concludes, for example, that a Down syndrome baby is per se incapable of flourishing is demonstrating remarkably poor judgment. Any human faculty that can be used can be misused and judgment is no exception. Just as we do not dismiss reason simply because we sometimes make faulty inferences, so we should not dismiss judgment simply because we are sometimes prone to prejudice.[15]

To call for the use of judgment when considering the topic of disability is not to endorse any particular past judgment, nor to say that one judgment is just as good as another. It is rather to say that there are no absolute hard-and-fast rules that can be applied to this issue that provide satisfactory solutions across the board. If we are to deal adequately with the wide variety of cases that fall under the umbrella of disability we are going to have to actually think for ourselves about specific cases. Sometimes we have to do philosophy without a net.

8.3 Paradigm Cases

8.3.1 Helen Keller as a Paradigm Case of Flourishing

Let's turn now to see how we apply *phronesis* in some paradigm cases, with attention to the theoretical advantages that follow as a result. Our starting question is "given the individual circumstances of this person's life, are they living well, or living poorly?" Aristotle believed that by and large disability made it impossible to live well; it is not hard to find clear counter-examples to this position. One of the most illustrative cases would be that of Helen Keller. Although her story is well known it will be worth our while to reiterate a few relevant details. After contracting scarlet fever when she was 19 months old, Keller was left blind and deaf. The story of her youth and early education at the hands (literally) of her remarkable teacher Anne Sullivan were famously chronicled in the play and film *The Miracle Worker* (Penn & Coe, 1962). Omitted from those dramatic representations was the fact that Keller went on to learn to speak, and to read Braille not only for English, but also for French, German, Greek and Latin. She attended Radcliffe College and became the first Deafblind person to earn a Bachelor of Arts degree. In addition to being a world-famous speaker, author and advocate for people with disabilities, she was also a staunch advocate for women's suffrage, birth control and pacifism, not to mention a co-founder of the American Civil Liberties Union.

The amazing accomplishments of Keller's life are all the more remarkable because her achievements were made, not so much in spite of her disabilities, but rather (at least in part) because of them. Given the circumstances of her life, what sort of judgments should we make regarding how well she lived? Keller's story is a paragon of several of the key virtues that we commonly think constitute a good life: courage, strength, resiliency, self-knowledge, compassion and wisdom, to name but a few. Her life is a shining example of the best attributes of the human spirit. It was certainly a life filled with difficulties that most of us do not have to deal with, yet that struggle elevated rather than hindered her. When we think about what kind of persons we want our children to be, what kind of life we want them to live, we could do much worse than using Keller's life as an exemplar.

Would Keller's life have been better or worse had she not been blind and deaf? Who can say? But more to the point, what difference does it make? It would certainly have been profoundly different, so different in fact that it seems fair to say she would have been a different person entirely. That hypothetical person's life may or may not have been as eudemonistic as the real Helen Keller's, but the measure of a person's life cannot be determined by contrast with the whole array of lives that they might have lived.[16] (If it were, then is there any among us that could truly say with confidence that they measure up?) To make judgments about a person's life we do not need to speculate on complex (perhaps unanswerable) counter-factual questions involving intricate metaphysical conundrums. Keller lived her life very, very well; at the end of the day, that is all the matters.

Not everyone with disabilities lives such a praiseworthy life, of course. But then again, most people without disabilities do not live such lives, either. Helen Keller's is not simply an exemplary life for a person with disabilities, but exemplary for a person, period. The circumstances of her life were different than most people's, but the virtues she embodied were among the same ones that ennoble all our lives. The case of Helen Keller shows us both how judgment can be employed in a particular case to assess individual flourishing, and also how disability can enable, rather than inhibit flourishing.

This is not to say that disability never inhibits flourishing. As with any circumstance in life, different people react to disability in different ways. It is not hard to imagine someone who, rather than drawing strength, endurance and wisdom from their circumstances, is instead overwhelmed by self-pity, anger and despair. A life marked most significantly by these reactions, scarred so deeply that they become not merely temporary emotional states, but rather a profound and fundamental aspect of the person's character, could not reasonably be said to be a life lived well. But yet again, there is nothing special about disability in this respect; someone may have a similar reaction to a personal scandal, the failure to succeed in his or her chosen career, or the death of a loved one. With a mind to the Stoic infusion that we gave to Aristotle above, Epictetus' reflections on disability are on point here: "Sickness is a hindrance to the body but not to the will, unless the will consent. Lameness is a hindrance to the leg, but not to the will. Say this to yourself at each event that happens, for you shall find that although it hinders something else it will not hinder you."[17] In each case what matters is much less the circumstances themselves and much more how we choose to respond to them.

8.3.2 Anencephaly as a Paradigm Case of Inability to Flourish

There are certain circumstances, however, that are so extreme as to make flourishing all but impossible. Anencephaly, a developmental disability in which a child is born without a cerebrum, is one such condition. Anencephalic children are born without the neural structures necessary to see, hear, feel pain or have any conscious awareness of themselves or their environment at all. There is no cure or standard treatment for children with anencephaly, and those that are not stillborn usually die from within a few hours to a few days after birth. It is, with little doubt, one of the worst imaginable conditions a newborn can have.

Could a child with anencephaly be said, in any meaningful sense, to "live well?" In assessing their lives we certainly cannot use any of the standard aretaic terms that apply to most people—these children are neither courageous nor cowardly, neither honest nor dishonest, neither kind nor cruel. There seem to be no terms of virtue or vice as we typically think of them that could apply here. What, then, would it mean for a child such as this to live well? Is there anything that we could realistically add to or take away from their lives to make them more eudemonistic? What could they do that would cultivate improvements of their character? These questions are nonsensical, of course, since by the very nature of their condition they do not have the requisite brain structures to have a character, to flourish, or to live well in any meaningful sense. In Epictetus' terms, anencephaly is a hindrance to the will, since it makes impossible any will, any consciousness, personality or character at all.

Perhaps the attempt to assess their flourishing in these terms is just a sublimated version of the species essentialism I derided above. According to this objection, just because "courage," etc. are the ways we measure whether or not people with a cerebellum live well, there is no reason to think that it is the proper measure for anencephalic children as well. They may have no will, but why insist that will is a necessary component for living well? If each life needs to be judged in terms of its own circumstances then we should consider anencephaly like any other disability and not presume that it inhibits flourishing, but rather ask how well the child is living given their condition. These children are alive, after all, and I previously cited the Aristotelian notion that anything that lives can live well or live poorly, so there must be some sense in which their lives can go better or go worse.

Now it seems we may be pushing up against the limitations of that Aristotelian notion. I certainly do not mean to merely assume that anencephalic children cannot have a meaningful life,[18] but I must admit to being unable to find any way around this conclusion. The brain is where we make meaning, where the experiences of our lives come together as a coherent narrative of value and significance. When that nexus fails to form (or is later destroyed) no meaning is possible. In the absence of any brain function, what could we possibly mean by the term "living well"? When looking at a given case we need to determine what aspects of their life are going well and which are going poorly. What would we identify in the case of an anencephalic child? Even if we discard our standard aretaic terms, are there *any* evaluative terms at all that can be applied? If someone wishes to say that such a child can live well or live poorly then the burden is on that person to direct our judgment towards the aspects of their life that are responsible for that difference.[19] In the absence of such an account we are compelled to conclude that it is impossible for anencephalic children to live well or have any kind of meaningful life.

It might be tempting for some to say at this point that anencephalic children have little to no "moral status." For virtue ethics, this determination seems thoroughly besides the point. In general, virtue ethicists tend to think that placing so much weight on the notion of moral status in these debates has lead us to lose sight of the larger picture.[20] The relevant questions with respect to these issues is not "what is the moral status of these children?" but rather "how do we assess their condition in light of the practical circumstances? What emotions, reactions, motives and actions are virtuous in these circumstances? What is the wise, compassionate, respectful, prudent, courageous, (etc.) thing to do?" These questions remind us of our role as agents and the role of our moral psychology, rather than myopically focusing on the status of the child.

Once we are reminded of the importance of moral psychology it is revealing to examine the extreme positions on the question of what we should do with respect to anencephalic infants. Those who do insist that an anencephalic child can live well or live poorly are likely to be the same as those who insist that we spend considerable resources to protect, prolong and increase the quality of their lives. Such people may seem to embody many virtues; a deep reverence for human life, a profound compassion for other beings and the integrity of their commitments and values. Given the medical facts in these cases, however, it seems that these character traits far surpass the virtuous mean and instead push into the vicious extremes; their reverence becomes close-mindedness, their compassion becomes over-sensitivity and their integrity becomes obstinacy. Part of what compounds the tragedy in these cases is how the emotional trauma of an anencephalic child can blind people to the medical realities at hand, leading them to make poor choices for all parties involved.

On the other hand, we must also avoid the other vicious extreme, in which one completely dismisses anencephalic infants. It does not follow from the fact that their capacity to flourish is drastically curtailed (to say the least), that therefore the lives of anencephalic are of no value, or that we may treat them however we wish. Were someone to suggest that these infants should simply be "put down" and anyone who thinks otherwise is "making a fuss over nothing" reveals mountains about their character. To hold such a position is to fail to appreciate the very same tragic nature of the circumstances mentioned a moment ago. Even if anencephalic children are incapable of living well the people who care about them surely are and they deserve our compassion, even if their grief has clouded their judgment. Anyone who fails to appreciate, in a deeply emotional fashion, the profound tragedy of the situation would prove themselves horribly callous and insensitive. While in my own judgment I cannot find a generally compelling reason to keep an anencephalic child alive, I do not see this as a simple or obvious solution, nor is it a conclusion I come to cavalierly.

Anencephaly makes flourishing impossible, but it is not the only condition in which this is the case. The considerations laid out here will obviously bear on similar conditions, including other cephalic disorders, patients in persistent vegetative states, and other cases of severe brain damage. I've chosen anencephaly here because it is as close to a paradigm case that we are likely to find. Nonetheless, this all-too brief sketch here does not provide us with an "all things considered" assessment of these similar cases. As I have repeatedly said, the various differences in the individual circumstances of these cases make such assessments very ill-advised. Every case must be considered in its own terms and thoughtfully judged. Nonetheless, looking this as a paradigm case can guide our thinking when it comes to the similar cases.

8.4 The Deliberate Creation of Disabilities

8.4.1 Creating Disabled Children

More complicated questions arise when we turn to judgments about the deliberate creation of disabilities in various contexts. One example of this occurs when certain would-be parents with disabilities, such as deafness or dwarfism, desire that their children take after them and pursue affirmative steps to increase the likelihood of that happening. These methods can involve anything from selective use of a sperm or egg donor, to pre-implantation genetic diagnosis (the genetic analysis of test-tube embryos prior to transfer into the woman's uterus). Several recent high-profile cases of this sort have provoked pointed reactions from critics and trenchant defenses from advocates.

Making moral assessments of these sorts of cases presents difficult problems for most moral theories because of the metaphysically complex issues of personal identity and harm to "future persons." Does the choice to deliberately bring a child with disabilities into this world constitute a harm to the child, or a violation of their rights? How could this be, given that at the time the decision was made the child does not yet exist? Perhaps the child that eventually comes into existence could later claim to have been harmed by their parents when they made their choice. But this solution is complicated by the fact that, had the parents not made that choice, that *particular* child would never have come into existence, and another, different child would have been born instead. How could the child charge that this choice harmed him when it was that choice that brought him into existence, such that he could be harmed in the first place?[21]

Virtue ethics can avoid these thorny issues (almost) entirely. As we saw in the last section, the core questions do not depend on issues of the status (moral or metaphysical) of the other person. Rather, they focus on the agent(s) in question, in this case the parent(s) deciding to conceive the child. We need to ask "what does it say about this person that they would deliberately bring a child with this particular condition into the world?" This question will not, by and large, turn on any perplexing metaphysical questions about future persons.

I suspect that when this question is asked in the abstract, many people would take a rather harsh view of such parents. They envision the life of a person with a disability as one of considerable physical and social hardships, coupled with the deprivation of the most meaningful and pleasurable aesthetic, social and physical experiences. A parent who fails to prevent bringing such a child into the world is bad enough; a parent who actively *seeks* these burdens for their child seems much worse. This may seem to be the equivalent of child abuse, no different from breaking an infant's legs or gouging out a child's eyes in an attempt to purposefully impair one's own child. What could be more selfish, more reckless, more cruel? But such conclusions would be premature. We cannot expect to accurately make determinations of this sort in the abstract. The details and nuances of these sorts of cases must be looked at if we wish to make wise judgments about them. A close examination of one of these cases will show how applying *phronesis* yields conclusions that may seem counter intuitive to common-sense moral thinking about the nature of disability. With that in mind, let's consider a concrete example of just such a case.

8.4.2 The Case of Candice McCullough and Sharon Ducheneau

In 2002 the *Washington Post Magazine* published a profile piece on Candice McCullough and Sharon Ducheneau, a lesbian couple who had recently given birth to their second child. The focus of the piece was the couple's decision to try to increase the chances of having a baby who is deaf. Both women self-identify as members of the Deaf community[22] and wanted to have their child share this identity with them, so they decided, as they had with their first child, to use a sperm donor whom had been congenitally deaf. They were prepared to accept and love a hearing child, had that been the way things worked out. As Ducheneau put it, "A hearing baby would be a blessing. A deaf baby would be a special blessing."[23] After he was born, their son, Gauvin, was tested and determined to be deaf in one ear and to have severe hearing loss in the other—much to the relief of his parents.

Liza Mundy, the author of the article, does a wonderful job of painting an intimate portrait of the two parents and the reasons behind their decision. Mundy does justice to the controversial nature of the situation without letting it overwhelm the humanity of her subjects. By the end of the article, one thing seems quite clear: McCullough and Ducheneau, whatever their personal faults, cannot accurately be characterized as "selfish," or "reckless," much less "cruel." These are loving people who are committed to the well-being of their children and (despite a common intuitive reaction to the contrary) wish to do everything they can to make the best lives for them.

This conclusion might seem hard for some people to reconcile with their preconceived notion of deafness. The context of the decision makes all the difference in the world. Mundy's article suggests that McCullough and Ducheneau's motives are not selfish, nor are they exceptional. Indeed, their motives are pretty much the same as any parents-to-be: to create a cohesive family environment with a mutual cultural identity and to position themselves to be the best parents they possibly can be. As Mundy states, McCullough and Ducheneau "believe they can be better parents to a deaf child, if being a better parent means being better able to talk to your child, understand your child's emotions, guide your child's development, pay attention to your child's friendships."[24] Surely these desires and emotions evince, not a pair of callous, domineering monsters, but rather two caring, compassionate and loving parents. While a precise theory of "virtuous parenting" is beyond the scope of this chapter, I suspect any plausible general theory thereof would be hard pressed to exclude McCullough and Ducheneau based on this description.

At the same time, however, this is not the only description that needs to be considered. There are many basic facts that cannot be ignored. Barring later medical/technological intervention, Gauvin will never know the beauty of a Beethoven symphony, will never enjoy the sound of waves crashing on a beach, will never experience any of the myriad sublime sensations that only a hearing person can enjoy. He will face considerable difficulty living in a world designed by and for hearing people and, sadly, he will face discrimination, overt and oblique. Life for Gauvin will have obstacles and frustrations that a hearing child would not face, and this will be a direct result of his parents' deliberate actions. Certainly facts such as these have a bearing on an assessment of the character of McCullough and Ducheneau.

Yet what, precisely, do these facts say about their character? How does their character differ from that of (to use Mundy's example) "parents who try to have a girl[?] After all, girls can be discriminated against."[25] McCullough makes a similar parallel regarding race: "Black people have harder lives. Why shouldn't parents be able to go ahead and pick a black donor if that's what they want?"[26] Certainly if such parents chose a black donor *specifically* so that their child would suffer, or simply out of ignorance of the social difficulties that black people may face, we might call them cruel or reckless. Yet it seems highly unlikely that any actual parents make such a choice for that reason. People take steps to ensure that their children follow after them so that they "can feel related to that culture, bonded with that culture."[27]

The power of that cultural bond is something that many people take for granted, due to the fact that, for most parents the primary components of that bond (be they race, nationality, religion, family history, etc.) are available to them automatically without the need for extraordinary intervention (such as locating a sperm donor with the specific heritable traits in question). Yet any parent who has taken pains to ensure that their child is connected to their cultural tradition, especially if that tradition is not the dominant one permeating the world in which their child grows up, should be able to identify with what McCullough and Ducheneau are trying to do. Parents who take pains to connect their children to a culture enrich their children's lives (even when doing so places burdens on the child) by offering them a venue for personal identity and social capital that would otherwise be unavailable to them. Had McCullough and Ducheneau deliberately given birth to a hearing child they may have deprived him of a close, intimate connection with his parents and identification with a community that has a profound beauty and value. Either way, they would have placed a burden on their son. Hence, assessments of their characters cannot be primarily based on the fact that they deliberately placed a burden on their child. Rather, as is usually (if not always) the case, we must look to the psychology of the agents. In this case, the decision was made conscientiously, with due regard for both the burdens placed on their children and the value of what is received in exchange. Given the circumstances, it is all but impossible to be cavalier in casting aspersions of vice on McCullough and Ducheneau. Indeed, there is something noble, respectful, considerate and courageous in such a decision.

One might object here that the real issue is not the fact that the parents placed a burden on their child, but rather the severity and nature of that burden. Could the value of that personal/cultural identity and social capital in a deaf child's life truly outweigh the disvalue of the associated burdens? This sort of consequentialist accounting is precisely the sort of thinking that virtue theory tends to reject, for several reasons. First of all, as we saw when considering the case of Helen Keller above, it is incredibly difficult for anyone to make reasonable estimates of the values of the lives in question. Who can say they are (in Mill's terms) "equally acquainted with and equally capable of appreciating and enjoying both" [28] of these alternate lives? By their very nature, they are mutually exclusive; being acquainted with one precludes being acquainted with the other. Secondly, even if one could be adequately familiar with these lives, the heterogeneity of the values within them makes reduction to a common denominator impracticable. How many units of utility does one ascribe to a Deaf persons' sense of personal identity, or to the social connections they have in the Deaf community? And how does that stack up against the utility of classical music, or the beauteous sounds of nature?[29]

More importantly, however, there is the myopia of such a homogenized approach to life in the first place. The knee-jerk reaction that burdening our children is per se a vicious thing, something that is to be avoided whenever possible, reflects an attitude that fails to appreciate how we, as human beings, actually live, grow and flourish. This is an attitude that Nietzsche recognized (and criticized) in the utilitarians; "What with all their might they would like to strive after is the universal green pasture happiness of the herd, with security, safety, comfort and an easier life for all [because] suffering itself they take for something that has to be *abolished*."[30] This "herd mentality," by depriving of us of our burdens, would in so doing deprive us of the challenges that we need to live well. Life isn't supposed to be easy and anyone who insists on making it so will find themselves unprepared to actually deal with adversity when it inevitably arises. That which does not kill us makes us stronger.[31]

The case of Helen Keller already showed us that such struggle can lead to flourishing, and we see it again here in looking at McCullough and Ducheneau. As Mundy puts it, both women "have faced obstacles, but they've survived. More than that, they've prevailed to become productive, self-supporting professionals."[32] Besides being well educated and successful (both have graduate degrees and work in the field of mental health), they are presented in the article as being loving, committed, courageous, independent and strong. If the worst that can be said about these parents is that they have condemned their children to a life of flourishing comparable to their own then it is hard to see what all the controversy is about.[33]

I want to reemphasize here the esoteric nature of cases such as these. It would be specious to conclude that because McCullough and Ducheneau display virtue, therefore in any case in which parents choose disability for their child, those parents are morally beyond reproach. It would likewise be a mistake to conclude that when a person loses their sight, hearing, use of their limbs or cognitive abilities, this is not a serious tragedy. The point in examining the case of McCullough and Ducheneau was not to answer all of these issues. Rather, it was to see how applying *phronesis* to a particular case yields conclusions that may run counter to mainstream thinking, to demonstrate that disability does not per se inhibit flourishing, and to show that the decision to have an impaired child is not necessarily a vicious one.[34]

8.4.3 Deliberately Avoiding Creating Disabled Children

Another consideration that is worth mentioning here involves prenatal testing to avoid giving birth to disabled children. Certain disability rights advocates have expressed concern at biotechnologies, such as pre-implantation genetic diagnosis, that allow parents to determine if their future child will be born with a physical, perceptual or cognitive disability.[35] The development of these technologies has been driven, at least in part, by the desire of parents to prevent having a child with such a disability. The objection is that these technologies, when used in this way, devalue persons with disabilities and deny them equal moral worth. If the primary uses of this technology were to prevent female babies or babies with darker skin tones, would anyone hesitate to call such uses sexist and racist?

When viewed through the lens of virtue ethics as I have construed it here, I think an equitable solution to this problem can be found. The basic problem arises when we look at the technology in isolation, without reference to the psychology of the people using it. As is so often the case, motives and emotions play a key role in assessing whether or not a given act is virtuous or vicious. Certainly if a couple used genetic technology to avoid having a female child because they believed that women were fundamentally inferior to men, then this would be inexcusably vicious. On the other hand, if they chose to have a male child for a more innocuous reason (say, because they already had daughters and they wanted at least one son) then it is hard to see any vice in this. Likewise, if a pair of prospective parents wants to avoid having a deaf child because they believe the life of a deaf child simply cannot have as meaningful an existence as a hearing child they are demonstrating their own ignorance and close-mindedness. By contrast, if those prospective parents wish for a hearing child because they believe they can be better parents to a hearing child (precisely the same reason McCullough and Ducheneau gave for wanting a Deaf child), while still willing to love and accept their child if it turns out to be deaf (or otherwise disabled), then we have no basis for claiming their actions are vicious.[36] If McCullough and Ducheneau's actions do not imply that hearing people's lives are of less value, or that they are not of equal moral worth, then how could the parallel, yet opposite actions of a hearing couple have such implication for deaf people?[37]

8.4.4 Clitoridectomy and the Deliberate Disabling of Women

How far could this line of thought be taken? What about other cases in which a culture deems it desirable to physically alter their children for the sake of that culture? I have already implied that if the motives, emotions and material circumstances were similar to that of McCullough and Ducheneau's, then parents trying to have a blind child, or a child with dwarfism, could nonetheless be virtuous. Could the same line of thinking I am suggesting here be used to defend, for example, female genital mutilation? If this were happening in a culture where the procedure was deemed essential for social identity and cohesion, where the act was done with the proper motivations, proper medical treatment to minimize pain and risk, and so forth, could clitoridectomy[38] be considered virtuous?

This would certainly be a disturbing conclusion, to say the least; one could easily construct a *reductio ad absurdum* against my position on this basis. Thankfully, I do not believe that someone could mount a successful defense of clitoridectomy using the framework I've outlined here. Part of the reason for this is due to the fact that the principal justifications for clitoridectomy are based on demonstrably false beliefs. According to *The Encyclopedia of Bioethics*, the reasons for clitoridectomy include maintenance of cleanliness, maintenance of good health, enhancement of fertility and improvement of male sexual performance. All of these claims are medically unsubstantiated, or to put it more bluntly, simply false. Several of the social reasons given (preservation of virginity, prevention of promiscuity) are, while not scientifically falsified, nonetheless factually dubious. Other social reasons given (pursuance of aesthetics, increase of matrimonial opportunities) are of questionable social value and a tenuous (at best) connection to virtue.[39]

In light of these considerations we can see the irony in this supposed *reductio*; these cultures seem to believe that the non-mutilated woman per se has less chance at flourishing. This false belief runs precisely parallel to the common belief in Western societies that I have been attacking in this chapter—that persons with disabilities per se have less chance of flourishing. Indeed, in a culture in which female genital mutilation is the norm, a non-mutilated woman is a disabled woman. So rather than this being an instance where a condition is deliberately, yet virtuously imposed on a child, this is a case where the social construction of disability is all too evident. The proper remedy here is not a blind tolerance of the culture, but rather a transformation of it to the point where the condition in question is viewed as healthy and compatible with living well. This is precisely what disability activists have been trying to do, only with respect to people with disabilities, rather than with respect to women.[40]

As a closing thought, it's worth mentioning that the language of virtue ethics provides us a rich pallet of opprobrium with which to condemn the practice of clitoridectomy and those who practice it. To describe this practice as "utility sub-optimal" or "a violation of the categorical imperative" is to drastically understate the case. To treat women as though their sexuality were a threat to society is the very definition of *misogyny*; to mutilate women in order to exert control over them is *barbaric*; to coerce and intimidate women, to acquiesce and accept this practice, is *cruel, chauvinistic*, and *paternalistic*. The fact that the vocabulary of vice lends itself so naturally to the aspersions that most Westerners naturally cast on clitoridectomy and its proponents suggests an affinity for an aretaic approach to the issue and the moral ammunition it provides. Our default moral framework is that of virtue ethics.

8.5 Conclusion

Despite certain obvious tensions, Aristotle can provide us with insight into the moral problems associated with disability. To rehabilitate Aristotle, he needs to be divested of his species-essentialist biology, and we need to adjust his view of the role of judgment in assessing the individual circumstances of a person's life. Once we shift away from judging how an individual lives up to the species norm and toward judging the individual circumstances of a person's life on their own merits, new avenues open up for us. We can see that there is no direct relationship between disability and human flourishing, allowing us to accommodate standard intuitions about people like Helen Keller, while not having to abandon our intuitions about extreme cases, such as those of children with anencephaly. It also can challenge some preconceived notions about the moral value of deliberately having children with disabilities, while allowing us to make case-by-case assessments of parents who do so. While some related problems have been addressed, several of the major issues have been explicitly set aside, in particular those pertaining to social justice and public policy. I hope I have not implied that because disability can lead to flourishing, therefore society as a whole ought not to try to redress inequities associated with disabilities, such as problems of access, participation and the like. All of these issues, while related, are separate and need to be dealt with on their own terms.

Notes

1. "As to the exposure and rearing of children, let there be a law that no deformed child shall live." Aristotle, *Politics*, 7, 1335b.15 (Trans. Jowett, Benjamin).
2. I say "reputed to have claimed" because as often as this quote shows up in the literature on disability I have been unable to locate a specific reference from Aristotle's original writings, despite a considerable search.
3. See, for example, Buchanan, Brock, Daniels, and Wikler (2000, pp. 284–288).
4. Boorse (1997, p. 4).
5. Boorse is specifically trying to articulate a "naturalist" (as opposed to "normativist") theory of disease. "The classification of human states as healthy or diseased is an objective matter, to be read off the biological facts of nature without need of value judgments." (Boorse, 1997, p. 4) I suspect that Aristotle would agree that health and disease could be read off the facts of nature, but would be baffled by the suggestion that doing so does not involve value judgments.
6. Aristotle identifies health as one of the "goods of the body," not goods in the "highest and fullest sense" (which he reserves for "goods of the soul"), but goods nonetheless, important because they enable and allow for the exercise of goods of the soul. *Nicomachean Ethics*, Book 1, Sec. 12–15.
7. Whether or not Boorse would agree with Aristotle on this point I am not sure. He might simply reject Aristotle's contention here and side with the Stoics, holding that flourishing is entirely a matter of one's inner-life, rather than their external circumstances. Either way, I am not attempting to second-guess what Boorse would say on the relationship between health and flourishing; I am rather extrapolating what Boorse's view of health would imply for an Aristotelian view of flourishing.
8. In particular I think of such gems as "The female is, as it were, a mutilated male." Aristotle, *Generation of Animals* 2, 3: 737a27.
9. We will see the application of this question to concrete cases in Part III.
10. It is only fair to point out that not everyone believes that Aristotle's original theories couldn't capture this in its own right. In particular, D. M. Balme argues in his article "Aristotle's Biology is not Essentialist" that Aristotle's view of species flourishing is sensitive enough to circumstances to account for this concern. According to Balme, Aristotle "argues for teleology with the question 'What benefits an animal of this kind?', not with the question 'What benefits *all* animals of this kind?'" He treats species as merely a universal obtained by generalization. While it is true that species-membership may help to explain the features of individuals, this is not because species is an efficient cause of individual formation, but because individuals in like circumstances are advantaged by like features." (Balme, 1987, p. 291). Whether or not Balme's interpretation of Aristotle's intent is correct or not, this is certainly a very plausible rendering of the underlying ideas, one with which I concur.
11. With regard to questions of public policy or social justice I do not doubt that the biological/social debate may be of tremendous importance. Yet because that discussion, from a virtue-theoretic point of view, would need to be informed primarily by a concept of human flourishing, that discussion needs to take a back seat to the current one. Hence, as I mentioned in the introduction, questions of public policy and social justice will not be addressed here.
12. Since I contend that the distinction will not be relevant to my purposes here I will generally continue to use the term "disability" to cover both the social and the biological considerations, for the sake of simplicity.

13. See, for example, Aristotle, *Nicomachean Ethics*, 1142a; or more recently, Hursthouse (1999) and Foot (2002).
14. Furthermore, it assumes that those who are critically analyzing it will be honest with respect to how they judge the cases at hand. It is fairly easy for philosophers to imagine hypothetical interlocutors who fail to share the intuitions driving an argument. While this form of critical analysis is both fair and useful it should not make us lose sight of the fact that we, as individuals, may actually assent to the judgment in question. As Rosalind Hursthouse says, "We *could* object, theoretically, to their moral premises and point out that. . . someone from a different cultural background might disagree. But do *I* disagree? If I do not, then their [virtue ethicists'] arguments are relevant to me and I should take them seriously" (Hursthouse, 2000, p. 155).
15. A complete defense of judgment would perhaps require a discussion of what constitutes good judgment as opposed to bad, how we can distinguish the two, how to correct for bad judgment, etc. As this article is not intended to be a complete defense of judgment these fundamental questions must be set aside for another date.
16. I do not mean to suggest that counterfactuals have no role in evaluating our lives. Certainly I can make some reasonable claims regarding how my life could have, in all likelihood, gone better, if certain facts had been different. My point here is simply to say that there are limits to how far such comparisons can (and should be allowed) to go. When the facts in question pertain not merely to ancillary aspects of my life, but instead constitute, in the deep sense, who I am—my character—then at that point such comparisons cease to be meaningful; we are then comparing two completely different lives, not two different versions of the same life.
17. Epictetus (1966, p. 135, paragraphs 8–9).
18. I should note here that by the phrase "meaningful life" I mean a life that is *meaningful for* the child in question. This is to distinguish it from a life that is *meaningful to* a third party, such as that child's parents, family, or even society at large. A full accounting of these sorts of cases would need to take into account the meaning these children have to such third parties, but in the current context I am setting such considerations aside for the sake of brevity and simplicity.
19. The only contender, it seems to me, are the basic autonomic and regulatory bodily functions— the beating of the heart, the inhalation of the lungs and so forth. These are the functions that Aristotle identified as the "vegetative" part of the soul (*De Anima*, III, 12, 434a 22–26), a notion that persists in the term "persistent vegetative state." While the term is considered by some to be insensitive due to its explicit dehumanization of the subject, it is hard to deny its analogical accuracy. If the only way in which anencephalic children can be said to flourish is in terms of mere biology then, in this respect, there is nothing that differentiates them from vegetation. There may be other respects in which their lives will differ from vegetation; the fact that we may have a profound emotional investment in our newborn children, anencephalic or not, would be one such difference. The key point here, however, is that due to their condition our emotional investment in them does not, in itself, make an anencephalic child's life more eudemonistic. Inasmuch as they can live well or live poorly, they do so irrespective of our feelings towards them.
20. Rosalind Hursthouse made this point with respect to abortion, claiming, "the status of the foetus—that issue over which so much ink has been spilt—is, according to virtue theory, simply not relevant to the rightness or wrongness of abortion. . . Or rather, since that is clearly too radical a conclusion, it is in a sense relevant, but only in the sense that the familiar biological facts are relevant" (Hursthouse, 1997, p. 228). In developing this line of thought, I follow Hursthouse's lead, applying her perspective on abortion to the case at hand.
21. Derek Parfit's seminal *Reasons and Persons* (1986) is still the standard text for outlining these sorts of issues with thoughtful detail and a profound respect for the complexity of the problems, even though it lacks clear solutions to them.
22. The capital "D" in "Deaf" here is meant to signify not merely the physical condition of being unable to hear, but rather a personal and social identity oriented around Deaf culture.
23. Mundy (2003, p. 70).

24. Mundy (2003, p. 73).
25. Mundy (2003, p. 73).
26. Mundy (2003, p. 73).
27. Mundy (2003, p. 73).
28. Mill (2001, p. 9).
29. There are, of course, sophisticated and respectable answers to these questions that a consequentialist could give. This is not meant to be an exhaustive argument against a consequentialist take on these issues. Rather, this is a cursory sketch of some problem areas in which the virtue theorist would press the consequentialist.
30. Nietzsche (1990, p. 72).
31. I am grateful to Susan Stocker and her insightful article "Facing Disability with Resources from Aristotle and Nietzsche" (2002) for drawing my attention to this application of Nietzsche's thought.
32. Mundy (2003, p. 73).
33. With this in mind, it is worth noting that despite his ill opinion of disability, Aristotle might inadvertently give support to McCullough and Ducheneau when he says, "anyone who does not take after his parents is really in a way a monstrosity" (Aristotle, *Generation of Animals*, IV, 3, 767b 5). This is surely taking Aristotle out of context, but in so doing a potent point is made: all parents want a child who takes after them in some respect or another; McCullough and Ducheneau are no exception. And while the shoe is on the other foot, it is also worth noting that Nietzsche might be harshly critical of the couple; one of his famous aphorisms asserts, "Without music life would be a mistake" (Nietzsche, 1998, # 33).
34. What would we make of a case similar to McCullough's and Ducheneau's, with a few differences: despite their preconception efforts, the baby is born with full hearing, at which point the parents took affirmative measures to make their baby deaf? What if, for example, they gave the child a (hypothetical) drug that destroyed his hearing, or surgically altered his eardrums to render them inoperable? How much of a difference does the timing (prenatal, in utero, neonatal), or the method (donor selection, genetic engineering, pre-implantation genetic diagnosis, pharmacology, surgery) make? I cannot speak to all of these questions here, so I will have to content myself with two basic points. First, several of these questions run parallel to problematic issues concerning the topic of abortion, and a virtue-theoretic approach to that issue, such as Rosalind Hursthouse's pioneering "Virtue Ethics and Abortion" (1997) may help illuminate these problems. Second, each of these particular cases would have to be met with close consideration of timing, method, motivation, material and cultural circumstances, and so forth, bringing our best moral judgment to bear on the details before any conclusions could be reached.
35. See, for example, Amundson and Tresky (2007, pp. 541–561).
36. To reiterate, for the record, proper motive alone is not the only relevant factor. Other elements, such as proper emotions and material and cultural circumstances are also relevant. I leave these out of the current discussion for the sake of simplicity.
37. One possible reply to this question appeals to the existing power structures in society, pointing out that since deaf people are a decided minority and hearing people are a decided majority that in this context, motives are besides the point. While I am not insensitive to this line of criticism, fully addressing it would require a much broader and more extensive discussion involving topics such as social justice, which I have deliberately avoided here. I do not have room to consider these topics here, and as such, I must leave them for another time.
38. Strictly speaking, clitoridectomy is merely one of four major varieties of female genital mutilation, as classified by the World Health Organization (see RHR & WHO, 2008). For the sake of simplicity, I will continue to use the term "clitoridectomy" as a generic term for any and all types of female genital mutilation.
39. Post (2004, pp. 413–417).
40. This conclusion should come as no surprise, as the disability rights movement has specifically tried to model itself on the women's-rights movements and other civil rights struggles. For more on the relationship between the disability rights movement and the civil rights

movement, particularly as expressed in legislative endeavors such as the Americans with Disabilities Act, see Burgdorf (2006). For some representative 'feminist' approaches to disability, see Wendell (1989) and Morris (1993).

References

Amundson, R., & Tresky, S. (2007). On a bioethical challenge to disability rights. *Journal of Medicine and Philosophy, 32*, 541–561.

Aristotle. (1885). *Politics* (J. Benjamin, Trans.). New York: Oxford University Press.

Aristotle. (1943). *Generation of animals* (A. L. Peck, Trans.). New York: Harvard University Press.

Aristotle. (1986). *De Anima* (H. Lawson-Tancred, Trans.). New York: Penguin Classics.

Aristotle. (1999). *Nicomachean ethics* (M. Ostwald, Trans.). Upper Saddle River, NJ: Prentice Hall.

Balme, D. M. (1987). Aristotle's biology was not essentialist. In A. Gotthelf & J. Lennox (Eds.), *Philosophical issues in Aristotle's biology* (pp. 291–312). New York: Cambridge University Press.

Boorse, C. (1997). A rebuttal on health. In J. M. Humber & R. F. Almeder (Eds.), *What is disease?* (pp. 1–134). Totowa, NJ: Humana Press.

Buchanan, A. E., Brock, D. W., Daniels, N., & Wikler, D. (2000). *From chance to choice: Genetics and justice*. New York: Cambridge University Press.

Burgdorf, R. L., Jr. (2006). Americans with Disabilities Act of 1990 (US). In G. L. Albrecht (Ed.), *Encyclopedia of disability* (pp. 93–100). Thousand Oaks, CA: Sage Publications, Inc.

Epictetus. (1966). The manual of Epictetus. In J. L. Saunders (Ed.), *Greek and Roman philosophy after Aristotle* (p. 135). New York: The Free Press.

Foot, P. (2002). *Virtues and vices and other essays in moral philosophy*. Oxford, UK: Oxford University Press.

Hursthouse, R. (1997). Virtue theory and abortion. In R. Crisp & M. Slote (Eds.), *Virtue ethics* (pp. 217–238). New York: Oxford University Press.

Hursthouse, R. (1999). *On virtue ethics*. Oxford, UK: Oxford University Press.

Hursthouse, R. (2000). *Ethics, humans and other animals: An introduction with readings*. London: Routledge.

Mill, J. S. (2001). *Utilitarianism* (2nd ed., G. Sher (Ed.)). Indianapolis, IN: Hackett Publishing Company, Inc.

Morris, J. (1993). Feminism and disability. *Feminist Review, 43*, 57–70.

Mundy, L. (2003). A world of their own. In O. Sacks (Ed.), *Best American science writing 2003* (pp. 68–87). New York: Harper Collins.

Nietzsche, F. (1990). *Beyond good and evil*. R. J. Hollingdale (Trans.). New York: Penguin Books.

Nietzsche, F. (1998). *Twilight of the idols*. D. Large (Trans.). New York: Oxford University Press.

Parfit, D. (1986). *Reasons and persons*. New York: Oxford University Press.

Penn, A. (Director) & Coe, F. (Producer). (1962). *The miracle worker* [Motion Picture]. United States: Metro-Goldwyn-Mayer.

Post, S. G. (Ed). (2004). *Encyclopedia of bioethics* (3rd ed.). New York: Thompson Gale.

Stocker, S. (2002). Facing disability with resources from Aristotle and Nietzsche. In W. J. M. Dekkers (Ed.), *Medicine, health care and philosophy* (Vol. 5, pp. 137–146). Dordrecht, The Netherlands: Kluwer Academic Publishers.

RHR & WHO. (2008). *Eliminating female genital mutilation—An interagency statement*. Geneva, Switzerland: Department of Reproductive Health and Research (RHR) & World Health Organization (WHO).

Wendell, S. (1989). Toward a feminist theory of disability. *Hypatia, 4*, 104–124.

Part III
Disability, Social Justice, and Public Policy

Chapter 9
Equal Treatment for Disabled Persons: The Case of Organ Transplantation

Robert M. Veatch

Mill is famous for his clever preliminary solution to the conflict between justice and utility-maximizing. In Chapter 5 of *Utilitarianism* he notes that, because of declining marginal utility, one can often maximize utility by concentrating resources on the worst-off in society, thus providing a utilitarian explanation for our intuition that the worst off have special moral claims. In medical ethics, reflection on cases of persons with severe, chronic disabilities challenges this felicific, if fallacious, resolution (Mill, 1967, pp. 391–434). These persons can sometimes command enormous quantities of medical resources, but gain only marginal benefit from them. An organ transplant for someone who is worst off because she has only a short time to live would, for example, target the worst off, but produce only small benefit. Thus proposals for aggressive medical intervention to bring those with disabilities to levels of nearer normal health or restore them to that status provide a critical challenge to a theory of justice in the use of health resources.

The term *disability* is vague. It includes a wide range of organic and mental conditions. For purposes of this chapter, I will limit my attention to severe, chronic disabilities. These are the most challenging to a justice theorist, especially one committed as I am to an egalitarian perspective. I will further limit my focus to mental disabilities. I believe nothing turns on the distinction between mental and physical disability. They may, in fact, both be related to organic pathologies, particularly when they are congenital. To give additional focus to my discussion I will concentrate on the ethical problems of those with severe mental disability who are candidates for organ transplant.

9.1 Two Kinds of Justice Problems

9.1.1 Baby K: The Case of Total Unconsciousness

Two kinds of cases come to mind: impairments so severe that mental function loss is total and less-severe impairments involving significant, but not total, compromise

R.M. Veatch (✉)
Kennedy Institute of Ethics, Georgetown University, Washington, DC, USA
e-mail: veatchr@georgetown.edu

D.C. Ralston, J. Ho (eds.), *Philosophical Reflections on Disability*, Philosophy and Medicine 104, DOI 10.1007/978-90-481-2477-0_9,
© Springer Science+Business Media B.V. 2010

of mental function. Baby K illustrates one of these: a case involving total permanent unconsciousness (In the Matter of Baby K: 1993). Baby K was born with severe mental impairment—a condition known as anencephaly. Although the term literally means "without brain," in fact those diagnosed with this condition may have some lower brain function. Babies with anencephaly are permanently unconsciousness, and survival for more than hours or days is extremely rare even if the infant is born alive.

Baby K was born in the early 1990s to a woman who was passionately committed to preserving the child's life in spite of the impossibility of consciousness and the great likelihood of rapid death. Although most parents presented with such a baby apparently would choose to classify life-support as an "extraordinary"—that is, a disproportional means—and refuse consent to life-sustaining treatment, Baby K's mother insisted on such intervention. She did so in the face of the moral judgment on the part of clinical caregivers that such treatment would be of no value.

This case presents two potential problems of equitable allocation of resources. The one that received public attention was whether it was a just or equitable use of resources to devote time, money, and professional attention to delivering life-support for such a patient when others might gain more benefit and perhaps even be said to have a greater need. The second problem was whether an insurer should adopt policies that would provide for more equitable allocation of resources in the future.

9.1.1.1 The Demand for Respiratory Support in the Absence of Scarce Resource Concerns

As the case actually evolved, it presented at most an indirect problem of justice in the distribution of resources. The use of scarce resources by the hospital turned out to be a non-issue. Personnel at the hospital claimed that they had enough professional staff and neonatal intensive care beds to provide the treatments for Baby K without jeopardizing the welfare of others. Moreover, the bills were being paid by a private insurer ready, willing, and able to make the payments. Hence, from the point of view of the hospital, there was no scarce resource problem. In part because of the lack of any evidence of resource scarcity and in part because of a recognition that Baby K's mother was making a value judgment about which she had a special claim of responsibility when caring for her daughter, the courts recognized her right to obtain emergency resuscitation from the hospital personnel.

9.1.1.2 The Hypothetical Concern About Scarce Resources

Although the specific facts of Baby K's case led the courts to siding with the mother in granting access to the service she demanded (and almost all other similar cases have been decided similarly), the case came close to presenting much more controversial issues. Had the insurer not been willing to pay the full costs of Baby K's treatment, the hospital would have faced the question of whether to use other resources. This would have posed the question of whether using common assets to fund Baby K's life-support was fair. Other patients in need might have had to go without treatment in that case.

In fact, the problem was not entirely hypothetical. The insurer should have faced the question of whether it was just to use the pooled assets of the subscribers to fund this treatment rather than to pay for benefits for others. Of course, the insurer might have decided to limit funding and simply pocket the savings. The morally appropriate way to conceptualize the insurer's situation, however, would be to view insurers as something analogous to public commodities with a rate set by a public agency and a certain portion earmarked for "profit," or in the case of a non-profit insurer (such as the one funding Baby K's treatment), for management costs and investment in future development. Another portion should be devoted to services for the pool of insureds. In this case, any payments for Baby K's treatment would come directly from the benefits for fellow members of the insurance pool. The critical question then is, what is a fair allocation?

In Baby K's case no one had thought in advance about whether a baby born with anencephaly should receive unlimited coverage for life support in the case when a parent insists. Since the insurer began paying without a clear policy in place, it is probably reasonable that they continued funding. To do otherwise would have cut the patient off in mid-treatment, after the decision-makers knew the social and economic status of the patient. Ad hoc bedside rationing decisions are certain to be controversial and could lead to charges of unfairness.

In the future, however, insurers should face the question of whether funding from the insurance pool of expensive treatments for those with severe disabilities is warranted. Presumably, some treatments will be warranted. After all, such coverage for seriously afflicted patients with desperate needs is what insurance is designed to provide. However, not every treatment for every disabled person deserves coverage. Baby K's case poses the problem of the criteria for limiting funding.

Some treatments for the severely and chronically disabled will be judged by typical members of the insurance pool to be of no value. These should be the first and most obvious treatments to be limited. For example, some expensive treatments for severely afflicted patients have been shown by peer reviewed outcomes research to have no effect on the disability. If a treatment is objectively demonstrated to be without any effect on the condition of the disabled, surely that is an obvious candidate for exclusion from insurance funding.

The problem is that in some cases a minority of subscribers to the insurance may sincerely and passionately believe that the treatment is effective even in the face of strong scientific evidence to the contrary. These patients and their loved ones will be seriously distraught if they are told the treatment is unavailable because the insurer will not pay for it.

In spite of the distress for this minority of patients and loved ones who will not get their desired treatments funded at the expense of the insurer, it seems clearly to be justified for the insurer to exclude such coverage so that limited resources can be devoted to treatments deemed effective. A clever insurer might realize at this point that the exclusion will alienate a minority of its subscribers. There would be no ethical reason for the insurer at that point not to offer an "insurance rider" so that those who persist in valuing the treatment deemed ineffective could pool their resources, buy the supplemental insurance rider, and set up a separate funding pool for the treatment seen as effective by only a minority. For example, a minority with a belief in alternative therapies might want to create a voluntary pool of funds by purchasing an insurance rider for their health policies. Fairness, however, requires that the main group of insureds not bear the cost of paying for a treatment judged obviously ineffective by the majority (Veatch, 1997, pp. 153–169).

Baby K's case is somewhat different. All involved were forced to acknowledge that the treatment demanded by the mother was effective in achieving what the mother sought: maintenance of respiratory function sufficient to preserve life (at least temporarily) even if the treatment was acknowledged not to offer the possibility of creating brain functions that could support any level of consciousness. Thus if the insurer wanted to exclude long-term ventilatory support for anencephalic infants such as Baby K, it could not do so on the grounds that the treatment was ineffective. It would have to do so on the more controversial grounds that the treatment, though effective, was not worth pursuing.

Claiming that life support for a severely disabled person is effective, but nevertheless not worth pursuing, is a controversial position. Normally, effective life-prolonging treatments for the disabled should be presumed to be worth pursuing. There may, however, be some limits on that presumption. In Baby K's case, a large majority of the population is in agreement that there is no recognizable benefit to providing life-support. The conservative Baby Doe regulations of the 1980s even made an exception for permanently unconscious patients (U.S. Department of Health and Human Services, 1985, pp. 14878–14892). Thus, it seems likely that a large majority of the subscribers to an insurance plan would, if asked, claim that such life-support is of literally no value and should not be funded. The Catholic Church would support such a conclusion, calling such treatment an "extraordinary" means, that is, a means that does not offer benefit.

Actually, a significant minority of the population appears to believe that anencephalic infants and others in a permanent coma or permanent vegetative state are actually deceased (Siminoff, Burant, & Youngner, 2004, pp. 2325–2334). Since normally there is no obligation of an insurance company to fund continued support of dead bodies, this group would presumably automatically oppose funding of the treatment.

I see no reason why a group of insured persons could not make a judgment that some possible treatment was effective in preserving life, but of no value. Certainly, they should not be expected to fund treatments on bodies they classify as deceased. Not only would persons be within their rights if they insisted on exclusion from insurance coverage of treatments judged by objective evidence to be ineffective; they would also be justified in insisting on exclusion from coverage of treatments judged by the majority to be of no value even though they were effective in preserving life.

This does not mean that such treatments would never be provided. I have already suggested that the minority who see value in these treatments could band together and create a supplemental insurance pool to fund the treatments. They might also be funded by private charities created by groups who see value in these treatments. Jewish scholars, for example, sometimes favor life-supporting interventions for patients when those interventions are deemed to be without value by the larger community. If a synagogue chose to pool resources and create a charity for supporting such life-sustaining treatment, no injustice would be done as long as the funds were private and were obtained licitly. Presumably, patients and their families could also self-fund such treatments if they were within their means.

Thus, ethics would permit the exclusion from insurance coverage of certain interventions either because they were objectively demonstrated to be without effect or even if effective were deemed of no value. Still, such treatments would not be banned. To the contrary, the American courts have rightfully concluded on several occasions that persons have a right of access to these treatments even against the judgment of professional caregivers, provided there is a private funding source and no other burdens are generated on other patients (In Re The Conservatorship of Helga Wanglie, 1991; Rideout, et al. v. Hershey Medical Center, 1995; Velez v. Bethune et al., 1995; In the Matter of Baby K, 1993; In Re Jane Doe, 1991). As long as there is sufficient staff and equipment to provide the treatment and it is equitably funded from private sources, there is no good reason to prohibit its provision. Respect for the minority who see value in such treatments requires delivery of such treatments in these cases.

9.1.1.3 Baby K and Organ Transplant

Those contemplating Baby K's case anticipated that an additional problem of resource allocation could have arisen. As the baby received highly skilled and competent life support, she did not die as predicted. In fact, she lived for two-and-a-half years until she succumbed to an infection. During that time there were concerns that her vital organs—kidneys and liver—might fail and that her mother might want an organ transplant to preserve her baby's life. Fortunately, the case did not escalate to that level of intervention, but had it done so, the community would have had to face the question of whether Baby K should be listed for an organ transplant just like any other patient with vital organ failure.

Had that question arisen, many—perhaps almost everyone—would accept that the transplant was not warranted. Perhaps even the parental surrogate would reach that conclusion. I see no moral problems if the surrogate were to conclude that, even though life-support was being provided for the severely disabled patient, a transplant would present risks so great that it would offer more expected harm than benefit. The surrogate surely would have the right to forgo the transplant as an extraordinary or disproportionally burdensome treatment.

The interesting question, however, is what should happen if the surrogate concluded that the transplant was worth pursuing. In other similar cases involving adults the patient himself or herself might actually have formed a conviction that he or she would desperately want to be transplanted if it would preserve life even in a permanent coma or vegetative state. Theoretically, such wishes could be documented in an advance directive, leaving the community with a case of a patient demanding a scarce life-prolonging organ even though permanently disabled to the point of unconsciousness.

If a patient or patient-surrogate for a severely disabled patient insists on being listed for an organ transplant, it poses an unexpectedly complicated problem for ethical theorists. The problem arises because most theories of justice require arranging social practices (such as organ transplant) so as to benefit the worst off or to provide them with opportunities for a well-being more nearly that of others. If one assumes that people in need of an organ transplant are low on the scale of well-being and good candidates for being among the worst off, it would seem to give them high priority for organs, not grounds for excluding them from the transplant waiting list. In fact, the American transplant policy insists on allocating organs equitably as well as striving for efficiently producing good outcomes (United States Public Law, 1984) Insofar as justice is concerned, then, the disabled seem to have a *prima facie* claim for high priority for transplants. I wish to focus on the question of organ transplants for the mentally disabled as a kind of test case in this discussion of justice and disability.

Those holding utilitarian and other consequence-maximizing normative ethical theories might attempt to reconcile their intuitions against providing an organ for Baby K with their ethical theory by arguing that more good will predictably come from the organ if it is implanted in someone with a possibility of future consciousness. All those who claimed that no benefit would come from life-supporting ventilation for Baby K would plausibly claim that no good would come from an organ transplant either. Assuming that net good would be expected from transplanting available organs to those without the severe disability, a utilitarian would have the basis for arguing for a policy against listing those with permanent unconsciousness.

The utilitarian justification for excluding the permanently unconscious from transplant waiting lists faces all the problems associated with utilitarianism. In addition to being illegal to the extent that it violates the requirement that transplant policy strive to promote equity as well as utility, it has strongly counter-intuitive moral implications. We now have enough medical and sociological data to predict rather accurately which types of people will get the most benefit from an organ transplant. Historically, one gender did better with kidney transplants, so that a utility-maximizing policy would concentrate transplants primarily (perhaps exclusively) on males. At certain points in recent history certain racial groups have had predictably better outcomes. Elderly people have predictably short times to benefit from transplant and would plausibly be excluded. In short, excluding the disabled on grounds of relative lack of benefit seems to rest on a policy that would exclude most of us if it were carried consistently to its logical conclusion. We intentionally arrange our organ allocation policies so as to provide fair opportunity for all groups to obtain organs, not merely to maximize efficiency measured in predicted years of life to be gained from a transplant (Veatch, 1989, p. 1).

Both law and ethics, then, require a just or equitable allocation of organs. If Baby K and others who are permanently unconscious are to be excluded, some other grounds must be found (at least for those who are not excluded by their own advance directives or by surrogate decision).

Insofar as justice is concerned, the critical question is who is the worst-off group and how that group can be helped. Since the severely disabled are, by definition, very poorly off, they would seem to command the attention of those who would determine organ allocation policy based on this moral principle—that is, all who would follow current American law and policy.

While anyone needing an organ transplant might, at first, seem to be among the worst off, a more careful analysis may reveal some complications with that conclusion. First, we recognize that some people who are potential candidates for organ transplant do not have immediate need. For liver transplant, for example, all adult candidates are assigned MELD (Model for End-Stage Liver Disease) scores that predict the likelihood of death without a transplant. (Children are assigned analogous pediatric—PPELD—scores.) Those listed with low MELD scores will predictably do well for some time without transplant. The severely disabled with low MELD scores plausibly have no more or less of a claim to a liver than those with similarly low scores who lack the disability. This suggests a moral rule that will require more attention later: those with disability deserve the same moral priority for organs (and other significant medical technologies) as those with similar medical characteristics who lack the disability. This rule can, in cases of a low MELD or PELD score, mean that a severely disabled person would not be a high priority to receive an organ. They would be ranked low not because of their disability, but because of their relatively low need for the intervention such as an organ transplant.

What, however, of someone like Baby K at that point at which her need for an organ produces a very high score? Does the rule that those with disability deserve the same moral priority for organs as those with similar medical characteristics who lack the disability imply that even those with the most severe disabilities (such as permanent unconsciousness) deserve an organ if they have sufficiently high MELD or PELD scores? Not necessarily.

The rule we have stated is an example of the formal principle of justice—that equals should be treated equally. The difficulty, however, is in determining what counts as being equal. In the case of all the candidates for transplant with identical MELD scores we know that, to the best of our ability, they have an equal predicted time-to-death. That, however, does not establish that all persons with the same MELD score are equally poorly off. If different people with the same MELD or PELD score are not all equally poorly off, then they are not really equals and perhaps do not need to be treated equally.

One standard way of asking who is worse off would be to ask which life one would prefer to be assigned. In this case we can imagine two people, each of whom will die without an organ transplant. One will face death in a short time (for example, three months) without a transplant and will suffer all of the symptoms of acute liver failure in the months until death. The other person will also die in the same short time, but, because of unconsciousness, will experience none of the harmful effects of the liver failure.

It seems very plausible that most people, if forced to choose one of those terrible fates, would say that the one who is unconscious and does not suffer is actually better off than the one who will be conscious and suffer in the months until death. The point is not that transplanting the conscious person will leave him or her better off after the transplant than the permanently unconscious person would be. That would feed a utilitarian analysis. The concern of justice is for identifying the worst off person and adopting the policy that improves that person's lot. If the conscious person would be worse off dying without a transplant than the unconscious person, then the principle of justice would favor the transplant for the conscious one.

In effect, our rule of allocating organs (and other significant medical technologies) to the disabled equally with those who have similar medical characteristics and lack the disability does not require that our two candidates have the same priority. They really do not have the same medical characteristics. One is conscious and can experience suffering while the other is not. If being able to experience suffering is a relevant difference that makes the one with this capacity worse off, then the requirement of giving the organ to the worst off (among those who consent or have surrogates who consent on their behalf) supports a lower priority for the permanently unconscious. There is a condition that could be worse: facing death through months of conscious suffering. From this analysis grounded in the egalitarian interpretation of the principle of justice, the permanently unconscious patient such as Baby K would not have the highest priority for an organ. Such a patient should be treated the same way that other permanently unconscious persons are treated, which does not necessarily imply that they are among the worst off.

9.1.2 Moderate Chronic Disability: The Case of Trisomy-21

Now let us turn to the second kind of case: a chronic disability that is still severe, but nevertheless less of an impairment than anencephaly. I will continue to use mental disability as the example, but the arguments will be similar for physical disability. Let us consider the case of a person with Trisomy-21, in which the individual is sufficiently impaired such that he or she is judged not mentally competent to make medical decisions or manage his or her own affairs. Such a person could be living with family or in a group home with adequate supervision. Such persons are at risk for the same diseases as anyone else and are potentially candidates for an organ transplant. In fact, persons with Trisomy-21 may be at unusual risk for cardiac septal defects that could require a heart transplant. The question is how the chronic, significant, but not total disability should impact the allocation of organs for transplant.

There are some, perhaps working from a utilitarian perspective, who might attempt to claim that lives with such impairments are of lower quality. If the life saved by the transplant is of lower quality, that would provide a basis for a utilitarian to assign a lower priority for receiving an organ.

There are two problems with this position. First, it is hard to sustain the claim that the lives of persons with Trisomy-21 are of lower quality. There is no inherent suffering from the condition; in fact, some claim that the condition may spare one from the depression and anxiety of living a life with more normal mental function. Second, even if the quality of life judgment were true, the utilitarian position rests on the dubious moral foundation of the utility principle—that those who would get less benefit from an organ have a lower priority.

Our rule of organ allocation—that those with disability deserve the same moral priority for organs (and other significant medical technologies) as those with similar medical characteristics who lack the disability—seems, at least at first, to support the conclusion that in the case of persons who are suffering from equal degrees of organ system failure, they deserve the same consideration (qualified with the proviso that those who are conscious and will suffer without an organ are actually somewhat worse off than those who are permanently unconscious, and therefore the conscious group deserves a higher consideration). Once again, however, the analysis is complex. We need to explore whether the disability has any impact on the comparison and, if so, what the implications would be for organ allocation.

Current national organ allocation policy does not take into account in any way whether one has a disability—mental or physical—other than the organ system failure that is potentially to be treated with a transplant. In this sense, those with retardation and those with severe physical disability have claims to equal treatment.

The intellectual puzzle arises when we ask whether the disability may have an impact on other characteristics that are generally seen as legitimate grounds for allocating organs. One example poses the problem in a particularly critical way. Persons with an intellectual disability such as Trisomy-21 must meet the same listing criteria as anyone else who needs an organ. If they meet the listing criteria, they will receive their organ based on allocation algorithms that do not directly consider mental or physical disability.

One of the criteria often used for listing is the requirement that the patient have the ability to follow the complex treatment regimen, including anti-rejection medication and follow-up visits. For the severely impaired, this may turn out not to be a problem because they would be living with supervision under the direct attention of a parent or institutional caregiver who could supervise the taking of drugs and follow-up. Assuming that, with such support, the individual will be able to meet the criteria for taking medications and returning for follow-up, these persons should not fail to meet the criteria for listing.

The problem is more complex with the less severely impaired. Some adults with Trisomy-21 function at a high enough level that they can live more or less on their own in facilities that provide only modest supervision. It is possible for someone at this level of Trisomy-21 to be living more or less independently, but still not have the capacity to follow the complex regimen of medications to successfully survive the years following transplant.

Another criterion could potentially raise a similar problem. Presently, the number of years of predicted survival following transplant is not a direct criterion for allocating organs, but the general idea that those who have longer predicted survival following transplant deserve higher priority has been with us for a long time. In the 1980s we gave priority for kidneys to those who had better HLA match (a measure of tissue compatibility) (Takiff, Cook, Himaya, Mickey, & Terasaki, 1988, pp. 410–415; Connolly et al., 1996, pp. 709–714). To this day we give a priority for a perfect HLA match although, because of better anti-rejection medication, HLA is not as important a factor in predicting survival today.

Some disabilities—mental and physical—correlate with shorter predicted survival. Persons with trisomy-21, for example, have a life-expectancy that is about 20 years shorter than comparable persons without the chromosome problem. This poses a problem for the just allocation of organs. We seem firmly committed to the notion that persons with a disability deserve the same consideration for transplant as others similarly situated without the disability, but in some cases the disability itself changes conditions (such as ability to follow a regimen or predicted years of survival following transplant). Thus our principle of treating equals equally and the derivative rule that those with disability deserve the same moral priority for organs (and other significant medical technologies) as those with similar medical characteristics who lack the disability seem to beg the question of whether someone with a disability is relevantly similar to someone with the same organ system failure without the disability.

Our organ allocation system might, for example, use predicted survival as one of the criteria for allocating organs. If we are committed to refraining from taking disability directly into account in deciding who is worthy of high priority for an organ, yet we know that the disability affects predicted survival time, the disability might sneak in through the back door. It would, for example, if predicted survival time with a transplant were one of the criteria for organ allocation.

It seems irrational to hold that these indirect effects of disability should be ignored just because it is the disability that causes them. For example, if for some reason one had a disability, say a disability in the immune system, that made graft survival impossible but one were otherwise a plausible candidate for a transplant, it would make no sense to say that, since we are committed to equal treatment of the disabled, we should ignore all the effects of the disability in deciding whether this person should receive a transplant. In this hypothetical case the graft would be guaranteed to fail, and it would be guaranteed to fail solely because of the disability. Even one deeply committed to equal treatment must conclude that such a person should be excluded from the transplant waiting list. He or she should be excluded because the graft is guaranteed to fail—a well-accepted grounds for exclusion of any candidate. The fact that the guaranteed failure can be traced to a disability has to be irrelevant.

I am forced to the conclusion that our rule—those with disability deserve the same moral priority for organs (and other significant medical technologies) as those with similar medical characteristics who lack the disability—is compatible with taking the disability into account in deciding whether candidates with disabilities have "similar medical characteristics." Even if we are firmly committed to the idea that disability cannot exclude one from receiving a transplant or other medical procedure, we must accept the idea that disability may sometimes be relevant in deciding whether one meets the criteria for the procedure. If some procedure is administered only to people with more than one year to live and a patient who is otherwise qualified has a disability that will shorten his life so that he has less than a year to live, then that person must be excluded. Treating equals equally has, as a corollary, treating people with relevant differences differently even if the relevant difference is the result of a disability.

9.2 Justice and What is Morally Right for the Disabled

What has been said thus far seems to imply that disability may become an indirect factor in allocation of scarce resources, perhaps sometimes in ways that are counterintuitive. For example, it may create circumstances whereby a permanently unconscious infant is not the worst off candidate for a scarce resource and would get lower priority because of that disability. It might also make it impossible for a patient to follow a post-transplant regimen, thus disqualifying him or her for the transplant. It may reduce expected survival time below some threshold that would qualify a patient for a procedure. In this final section, I need to explore whether these indirect implications of disabilities are justified and, if so, why.

Our focus has been on the ethical principle of justice and its formal formulation—that equals should be treated equally and those with morally relevant differences differently. We now need to examine how this principle of justice and the rule we derive from it for purposes of allocating scarce medical resources should relate to other moral principles.

Almost all normative ethical theories include more than a principle of justice. They include a principle of respect for autonomy or a related principle of permission (Engelhardt, 1996). They often include consequence-maximizing principles of beneficence and nonmaleficence. If there is more than one principle in an ethical system, merely satisfying the principle of justice will not be sufficient. Our final moral judgment (or "duty proper." to use the Rossian term) must take into account the other ethical principles as well. Depending on our theory of resolution of conflict among principles, justice may lose out in the end (Veatch, 1995, pp. 199–218).

We have seen that current U.S. national policy requires simultaneous consideration of both justice and utility—equity and efficiency. The UNOS Ethics Committee has interpreted this to require that these factors be given equal weight (Burdick, Turcotte, & Veatch, 1992, pp. 2226–2235). Factors incorporated into an allocation formula usually can be attributed to one or the other of these considerations. (Occasionally, as in the case of extra priority for kidney allocation to children, both equity and efficiency can support the policy.) Likelihood of success and years of survival are, for example, incorporated to satisfy the principle of utility. They measure probabilistically the amount of good expected. Other considerations such as who is worst off and who has been waiting longest have no direct bearing on utility of an intervention. In fact, in medicine it is sometimes the case that those who have waited the longest have the worst chance of success from treatment. Nevertheless, these factors are morally relevant from the point of view of justice, especially egalitarian justice that focuses on who is worst off.

We have seen that, in the world of transplant, disability can contribute indirectly to criteria for allocation. Sometimes disability gives one lower priority because of a justice consideration—others who are conscious may be worse off. Sometimes disability gives one lower priority because of a utility consideration—those with disability get predictably less benefit. Likewise, it is conceivable that in some cases disability could indirectly make one worse off than others who are medically identical except for the disability. Mental impairment leading to inability to understand treatment might, for example, make one worse off. (Consider the justification for not providing chemotherapy to Joseph Saikewicz [Superintendent of Belchertown State School v. Saikewicz, 1977]). Similarly, a disability could lead indirectly to a procedure having greater utility. (Consider the benefit of a procedure to preserve sight in one eye of a person already blind in the other eye compared with the benefit of the procedure for someone sighted in the other eye.)

As long as national policy is committed to the legitimacy of considering both utility and justice equally in our allocation formula (for transplant or any other procedure), we probably have to live with the fact that disability may indirectly lower (or raise) the priority for either utilitarian or egalitarian reasons. Those who are uncomfortable with this conclusion may have additional reasons to reexamine the commitment to the legitimacy of giving utility and justice equal weight. It is striking that some of the most controversial implications of indirect consideration of disability are the cases in which the disability indirectly lowers the benefit of a procedure (by shortening the length of the benefit, for example). Those who are uncomfortable with this may need to examine whether utility should be given equal weight in allocation formulas. On the other hand the cases in which disability raises or lowers priority on justice grounds (such as the case in which a permanently unconscious anencephalic may have her claim for a transplant lowered) seems less controversial. If organs and other scarce resources are to be allocated according to the principles of egalitarian justice, the worst off will get priority even if it is their disability that makes them worst off.

Our conclusion is complex. The central principle in relating disability to organ allocation is that those with disability deserve the same moral priority for organs (and other significant medical technologies) as those with similar medical characteristics who lack the disability. That rule, however, is consistent with the possibility that in some cases disability can indirectly impact allocation priority by influencing whether those with disability have morally relevant medical characteristics. When they do not, the disability may indirectly influence organ allocation. Disability can make transplant more beneficial or less beneficial, a fact that would influence those who rely on utilitarian criteria of efficiency as at least one factor in allocating organs. Disability can also make persons worse or better off, giving the disabled higher or lower priority when the principle of justice is applied.

References

Burdick, J. F., Turcotte, J. G., & Veatch, R. M. (1992, October). General principles for allocating human organs and tissues. *Transplantation Proceedings, 24*(5), 2226–2235.

Connolly, J. K., Dyer, P. A., Martin, S., Parrott, N. R., Pearson, R. C., & Johnson, R. W. (1996, March 15). Importance of minimizing HLA-DR mismatch and cold preservation time in cadaveric renal transplantation. *Transplantation, 61*(5), 709–714.

Engelhardt, H. T., Jr. (1996). *The foundations of bioethics* (2nd ed.). New York: Oxford University Press.

In Re Jane Doe. (1991, October). A minor, civil Action File No. D-93064, Superior Court of Fulton County, State of Georgia.

In Re The Conservatorship of Helga Wanglie. (1991, June 28). State of Minnesota, District Court, Probate Court Division, County of Hennepin, Fourth Judicial District.

In the Matter of Baby K. (1993). 832 F. Supp. 1022 (E.D. Va.).

Mill, J. S. (1967). Utilitarianism. In A. I. Melden (Ed.), *Ethical theories: A book of readings* (pp. 391–434). Englewood Cliffs, New Jersey: Prentice-Hall, Inc.

Rideout, Administrator of Estate of Rideout, et al. v. Hershey Medical Center. (1995). Dauphin County Report, pp. 472–498.

Siminoff, L. A., Burant, C., & Youngner, S. J. (2004, December). Death and organ procurement: Public beliefs and attitudes. *Social Science and Medicine, 59*(11), 2325–2334.

Superintendent of Belchertown State School v. Saikewicz. (1977). 373 Mass. 728, 370 NE 2d 417.

Takiff, H., Cook, D. J., Himaya, N. S., Mickey, M. R., & Terasaki P. I. (1988). Dominant effect of histocompatibility on ten-year kidney transplant survival. *Transplantation, 45*, 410–415.

U.S. Department of Health and Human Services. (1985, April 15). Child abuse and neglect prevention and treatment program: Final rule: 45 CFR 1340. *Federal Register: Rules and Regulations, 50*(72), 14878–14892.

United States Public Law. (1984, October 19, 98–507). *National Organ Transplant Act* 98 Stat. 2339.

Veatch, R. M. (1989, July). Allocating organs by utilitarianism is seen as favoring whites over blacks. *Kennedy Institute of Ethics Newsletter, 3*(3), 1,3.

Veatch, R. M. (1995, September). Resolving conflict among principles: Ranking, balancing, and specifying. *Kennedy Institute of Ethics Journal, 5*, 199–218.

Veatch, R. M. (1997, June). Single payers and multiple lists: Must everyone get the same coverage in a universal health plan? *Kennedy Institute of Ethics Journal, 7*(2), 153–169.

Velez v. Bethune et al. (1995). 466 S.E., 2d 627 (Ga. App.).

Chapter 10
Disability Rights: Do We Really Mean It?

Ron Amundson

This chapter will argue that the disability rights (DR) movement has a much lower level of acceptance than other civil rights movements, especially within the academy. This is true even though the other movements are regarded (at least by disability rights advocates) as similar in nature. By "within the academy" I mean within the discourses, formal and informal, of professional academicians: professors and other intellectuals and their students. I will try to demonstrate by example that positions held by the DR movement are summarily rejected by many within the academy, even though similar positions are unquestioningly endorsed when stated by advocates of women's rights and "racial" or ethnic civil rights. My claim is not that women's rights and racial civil rights are genuinely supported within the academy. Racism and sexism still exist, and have serious negative effects. However, racism and sexism are almost never openly endorsed in today's academic discourse. The mismatch between discourse and practice is regrettable. However (I suggest) if practice matched discourse, minorities and women would have very nearly equal rights with majorities and men within the academy. Disabled people would still not have equal rights with nondisabled people. Basic DR principles are rejected not only in practice but also in discourse.

Within the academy, language that openly disparages groups that claim civil rights protection is almost never acceptable, and statements that challenge the civil rights goals of these groups are regarded as highly suspect.[1] I do not mean to celebrate the demise of sexism, racism, and heterosexism; the attitudes still lurk. However, the expression of such attitudes is now regarded as inappropriate at the very least. In contrast, I will argue, discourse that defends the justice of socially inflicted disadvantage to people with impairments is not only accepted within the academy—it is virtually the norm. I will include examples of anti-DR discourse from the media, and from producers of the media. Some of this material does not come precisely from the academy, but serves to illustrate the attitudes that are damaging to DR.

R. Amundson (✉)
Department of Philosophy, University of Hawaii, Hilo, Hawaii, USA
e-mail: ronald@hawaii.edu

D.C. Ralston, J. Ho (eds.), *Philosophical Reflections on Disability*, Philosophy and
Medicine 104, DOI 10.1007/978-90-481-2477-0_10,
© Springer Science+Business Media B.V. 2010

In order to pursue this discussion, some semantic stipulations are necessary. First, we must distinguish biomedical conditions from the disadvantages that might be associated with them. The purpose of this distinction is keep our attention on the contingent nature of the disadvantages that often accompany biomedical conditions. Unemployment, for example, is frequently associated with quadriplegia. But the DR movement rejects the "naturalness" of that association. It is no more natural to assume that a person with quadriplegia is unemployed than to assume that a woman or a member of an ethnic minority is unemployed.

The term *impairment* will refer to a biomedical condition that is presumed to be subtypical of the human race, without any assumptions about the disadvantages that might accrue to individuals who possess impairments. This is consistent with most DR literature. Disadvantage will be separated into two distinct kinds; those that are socially mediated (on the one hand), and those that are intrinsic to an impairment itself (on the other).

- *Conditional* disadvantages of impairment (CDIs) are disadvantages that are experienced by people with impairments, which are produced by the social context in which those people live.
- *Unconditional* disadvantages of impairment (UDI's) are disadvantages that are experienced by people with impairments, which are produced irrespective of their social context.[2]

The distinction between CDIs and UDIs is not completely a matter of objective measurement, of course. It can be politically contentious. For example, DR advocates might regard some disadvantages as socially caused that critics of the movement would regard as unconditional. But this happens in all civil rights movements. Opponents of the women's movement often claim that women's preponderant role as caregivers is a natural consequence of their (innate) psychology and biology; feminists consider it social discrimination. For our purposes, the argument need not be carried this far. The mere recognition of the CDI/UDI distinction will be enough to draw attention to the points we need to recognize. We are, after all, talking about the academy, an institution primarily made up of people who *believe that* they support DR, just as they (believe that) support the women's movement. Few of these people would claim that womens' roles as caregivers is amply justified by biological facts. Husbands among this group may not share childcare responsibilities. But if they don't, at least they have the grace to be embarrassed about it.

Socially disadvantaged groups are typically stigmatized (in the sense of stigma specified by Goffman, 1963). This chapter will illustrate the nature of the stigma as it continues to apply to people with impairments. That stigma is responsible for the continued failure of so many mainstream academics to come to terms with the DR movement, and to grant it the legitimacy that is granted to other civil rights movements. Stigmas are associated with *stigma theories*. These are ideologies: conceptual structures that rationalize particular beliefs about stigmatized groups, and make the disadvantages of these groups seem natural and inevitable. One aspect of the stigma of disability is the belief that UDIs (the disadvantages that are intrinsic to impairments) are immense, and that people with serious impairments are permanently disqualified from ordinary life by the immensity of their burden of UDIs. Such beliefs were commonly applied to women and minorities in past years. They continue to be applied to people with impairments today. Perhaps you doubt; let us see. The discussion will be partly historical, beginning with an illustration from the early days of the civil rights movement against racism.

10.1 Early Context

Case 1. The crippled veteran of the Pacific war tells my [African American] brother: 'resign yourself to your color as I got used to my stump; we are both casualties.' However, with all my being, I refuse this amputation. I feel like a soul as wide as the world, a truly deep soul, as deep as the deepest river; my chest has the power to expand infinitely. I am a don and I am advised to accept the humility of the crippled. (Frantz Fanon, 2001 [1951], p. 200, referring to a scene from *Home of the Brave*, a 1949 American movie directed by Mark Robson about racial tensions within an integrated group of American soldiers on a dangerous mission)

I ask the reader: Is this not a heart-wrenching plea for racial equality? But wait—save your answer until we discuss the circumstances in which Fanon refused his "amputation."

Frantz Fanon was among the most powerful anti-racist and anti-colonialist voices of the mid-20th century. Only the faintest glimmerings of a DR movement existed in 1951. The successes of the racial civil rights movement and the women's movement were needed before enough disabled people had the liberatory consciousness to forge a DR movement. Today's movement owes everything to voices like Fanon's, which revealed the assumptions behind the complacent racism of his day.

Half a century later Fanon's anguished plea should sound old fashioned, and even reactionary. Unfortunately (I suggest) it does not. I repeat the question: was that statement not a heart-wrenching plea for racial equality? To a DR activist the answer must be: No. It was an attempt to bargain for racial dignity at the expense of the dignity of people with physical impairments. I suspect that only a few readers would have recognized that fact immediately upon reading the passage. I confess that I (a DR activist) read it with a vague puzzlement two or three times before the implications of Fanon's statement came crashing home to me. Fanon insists that he is not like a cripple. He says that enforced humility is *not* deserved by people of African descent in the way that it *is* deserved by cripples.

The very power of Fanon's statement relies on the contrast between race and impairment. "Here's how badly white people think of us: they think of us as if we were cripples!" Such a comparison should not be assumed to be an insult by anyone who supports DR in the modern day, however much we may be able to forgive Fanon for his ableist ignorance. To compare a nondisabled person to "a cripple" should be no more insulting than to compare a white person to a nonwhite person, or a man to a woman. My purpose is not to criticize Fanon, a person of his times.[3] Nor do I mean to defend the movie-message that Fanon rejected. Racism and ableism are not the same, and it is quite understandable that Fanon should consider that message to be a trivialization of racism. But he could have defended the unique evil of racism without implying that amputees *genuinely deserve* the humiliation that he rejects for himself.

My purpose in this first example is to sensitize the reader (who may be unaware of the DR movement) to the difference between traditional views of disability and the views of modern DR advocates. Disability is regarded as neither shameful nor pitiable, but a fact of life that can be dealt with like any other fact of life. The comedian Jerry Lewis was a top movie star at the time Fanon's quotation was written. Lewis is now the spokesperson for the Muscular Dystrophy Telethon. He is also a regular target of DR protests and criticism because of his demeaning descriptions of the people who have the conditions his charity is intended to support. "Piss on Pity" is a frequent DR slogan of this protest. Lewis is angered by this criticism, because he considers himself a genuine humanitarian. He was asked about the protest on the CBS Morning Show in 2001, and his response was "If you're paraplegic and in a wheelchair and you don't want pity, stay in your house!" (CBS Morning Show, 2001).

Much of the world has moved past Lewis's old bigotries. Now that the environment is more accessible to wheelchairs, wheelchair use doesn't seem to arouse the pity that Lewis expects. Let us consider whether other aspects of the movement are equally understood and accepted.

10.2 Media

The media are widely recognized as expressing and promulgating stigmas of impairment, just as they have done in the past with the stigmas of race and sex. The surprising fact is the openness and self-consciousness with which this happens. Publications about the techniques of screenwriting, for example, do not openly advocate the use of sexist or racist stereotypes in order to propel story lines. But they do openly advocate the exploitation of the stigmas of impairment, in full recognition that they will be harmful to people who have the impairments being stigmatized. The following passage is written by Syd Field, widely regarded as one of the best teachers of screenwriting.

> **Case 2**. Pictures, or images, reveal aspects of character. In Robert Rosse's classic film *The Hustler*, a physical defect symbolizes an aspect of character. The girl played by Piper Laurie is a cripple; she walks with a limp. She is also an emotional cripple; she drinks too much, has no sense of aim or purpose in life. The physical limp underscores her emotional qualities— *visually*. Sam Peckinpah does this in *The Wild Bunch*. The character played by William Holden walks with a limp, the result of an abortive holdup some years before. Physical handicap—as an aspect of characterization—is a theatrical convention that extends far back into the past. One thinks of Richard III, or the use of consumption or VD that strike the characters in the dramas of O'Neill and Ibsen, respectively. (Field, 1994, pp. 31–32)

Note how Field describes his technique. Representations of a character's physical impairment are said to "reveal" a character flaw. Physical impairments represent character flaws as "pictures, or images... *visually*." This is utter nonsense. In the real world, one cannot learn about peoples' character flaws by observing their impairments. Physical impairments are in no way "pictures or images" of character flaws. Such notions are expressions of a stigma theory, an ideology. The stigma theory for physical impairment provides us with a code by which we can read character flaws out of impairments. But this "reading" comes entirely from the ideology, the bigotry, not from any associations that we have learned from the real world.

According to this ideology, everyone with a physical impairment either (a) is bitter and angry at the world because of their impairment, (b) got the impairment because of a character flaw and so somehow deserves it, or (c) in some other mystical, fairy tale-like manner possesses a character flaw that is symbolically reflected in their impairment. There is no truth in this fairy tale; it is pure stigma. Syd Field (the screenwriter) has not empirically discovered an association between physical impairments and character flaws. The ideological association in the media between impairment and character flaw is widely discussed in DR and related literature (Sontag, 1989; Wendell, 1996; Longmore, 1985; Norden, 1994; Darke, 1998). If Field were honest (and fully conscious of what he was doing) he would have said: "Physical impairment is popularly associated with character flaws. Like other social prejudices, this stigma can be usefully exploited in constructing screen plays." The ideological connection of character flaw with visible impairments does immeasurable harm to people with impairments. But this is of no interest to Field. He's a screenwriter, not a moralist.

Exploitation of racist and sexist stigmas are just as useful as ableist stigmas, and were openly used for many years. Civil rights movements have reduced this exploitation for most groups, but not yet for people with impairments. Field's book does not recommend using racist or sexist stereotypes as "an aspect of characterization" in the way that he uses ableist stereotypes. But if he had written in 1950 he may well have done so. (See the characterization of Zip Coon below.) We should learn from the fact that Field was still openly exploiting this stigma (but not those of sex and race) in 1994.

A direct and conscious exploitation of an ableist stigma can be seen in an interview of John Cleese, a comedian made famous by *Monty Python's Flying Circus* (a comedy series well-loved among philosophers). In an interview that aired in 2006. Cleese was asked to explain his talent at mocking authority figures. One answer was that he had developed a special understanding of how authority could be made to look ridiculous. He revealed his secret: authority figures should be depicted as disabled.

> **Case 3**. I realized very, very early on that if somebody, a character that you're going to write is, is, is, is going to do that, then it's funny if he's the head of the Secret Service and not funny if he's a milkman. So that the more authority that you give these characters, the more that they have hanging on them, the more people's lives depend on how they're going to act, then the f-f-f-funnier it is when they do a bit of that. (Cleese, 2006 [1986], emphasis added)

At this point of the interview, Cleese is sitting at the edge of his chair. During the underlined passages he tightens and twitches his head, neck, and right shoulder, and stammers to imitate the speech of a person with cerebral palsy. He begins to chuckle when he talks about the authority that his invented character has to affect the lives of others (perhaps being the head of the Secret Service) and at the end of the speech he collapses back into the chair laughing. He and the interviewer share a big laugh at the newly invented character: a director of the Secret Service whose speech and body movements are affected by cerebral palsy.

Notice the stigma that is being exploited in this passage. It is not a character flaw, but generalized incompetence. A person who is unable to fully control his body movements—who moves and speaks like a person with cerebral palsy—is incompetent to make decisions that affect the lives of others. This stigma is assumed to be so deep in the audience's mind that it need not even be spoken. If you twitch, you're incompetent—this is so obviously true that it is wildly funny to imagine you as being employed in a job with a high level of authority. Cleese doesn't try to associate the impairment with a particular demand of the job. The character with cerebral palsy is not assigned a job of, say, assembling watches (a job for which he may be unqualified because of his lack of fine motor skills). The person with cerebral palsy is depicted as ridiculous merely because of his presumed inability *to make important decisions*. Making important decisions does not require fine motor skills! But stigma overrides logic. This particular stigma is so deep that all that is needed to produce a hysterically funny character is to depict a person with cerebral palsy in a position of great responsibility.

If you're still chuckling at Cleese's character, try this thought experiment. Think of yourself as a well-educated person with cerebral palsy. Think of yourself as going out for a job interview. Now think of your interviewer as having recently seen and enjoyed Cleese's performance of his comedic character—the person in a position of great responsibility who has cerebral palsy. The job interviewer would do well not to laugh in your face. People with cerebral palsy (like people with many other categories of impairment) have a very low employment rate, even when they are well educated and are perfectly competent to do many jobs. Cleese is making a very good living by *exploiting and reinforcing* the stigma that keeps people with cerebral palsy unemployed. Philosophers love the guy.

A close analog to Cleese's exploitation of the stigma associated with cerebral palsy can be seen in the history of racist humor. During the early nineteenth century, minstrel shows traveled throughout the U.S. The first blackface character (i.e. a black character played by a white actor in makeup) was a slave called Sambo, who was depicted as lazy and ignorant. This served the interests of slave owners: the character showed why slaves did not deserve to be paid wages. Eventually, abolitionists began to make inroads on public opinion. Racist humor changed to meet the challenge. A minstrel show character was invented to mock the free blacks in the North, and ridicule their ambitions. The new character was named Zip Coon, a free black man who was "a dandy and a buffoon." Zip Coon's ridiculous attempts to imitate white people showed how futile the emancipation of blacks would be. A sample of Zip Coon's performance: "Transcendentalism is dat spiritual cognoscence ob psychological irrefragibility, connected wid conscientient ademtion ob incolumbient spirituality and etherialized connection ..." (Riggs, 1986).

Notice the similarities between the character of Zip Coon and Cleese's off-the-cuff invention of the Secret Service director with cerebral palsy. The black character was a member of a stigmatized group who was depicted as ridiculous when he tried to rise above his socially dictated station. Cleese's character of the Secret Service director with the impairment is exactly the same. In both instances, the audience roars with laughter at the absurd failures of both characters. This laughter performs the social service of strengthening the stigma. By the turn of the 20th century the musical category of what were called "coon songs" was among the most lucrative forms of entertainment, both in stage performance and sheet music publishing (Dormon, 1988). This was during the period of Jim Crow laws and lynchings. No one could question the connection between the "humor" of coon songs and the racist oppression of African Americans of this era. I submit that the same relation exists between Cleese's brand of humor and the social oppression of disabled people today. Just as Zip Coon shows that black people should not expect to be free, Cleese's head of the Secret Service shows that people with cerebral palsy should not expect to hold jobs of responsibility. The very low rate of employment of people with noticeable cerebral palsy, even those with advanced degrees, is a matter of public record (Henderson, 2006; Canadian Association, 2007). Like the composers and performers of coon songs, Cleese plays his part in enforcing this social arrangement.

10.3   Actual Arguments 1: "Disadvantages Remain"

From Chance to Choice (Buchanan, Brock, Daniels, and Wikler, 2000, henceforth FCC) is a very well-received book on the bioethics of genetic technology. It is also one of the first volumes on bioethics to take the DR movement seriously. The book's primary interest in the DR movement comes from the arguments of DR activists who have criticized modern genetic policy, claiming that it verges on eugenics. FCC discusses several of these arguments (with varying degrees of fairness) and rejects them all. My present concern is not with the rejection of the DR critiques, but with the characterization of the DR movement. FCC gives at least three distinct arguments to the conclusion that the DR movement has less moral legitimacy than other civil rights movements. The movements which are said to have more legitimacy than the DR movement include racial civil rights, women's rights, and gay rights. Here is one of those arguments.

Case 4a. The limitations a gay or black person suffers are injustices in a quite uncontroversial sense: they are forms of discrimination. While deaf people and others with disabilities certainly do continue to experience discrimination, *they would continue to suffer limited opportunities even if there were no discrimination against them.* . . .The fact that it is costly to remove barriers of discrimination against blacks or gays has no moral weight because no one can have a morally legitimate interest in preserving unjust arrangements. . . . the costs of changing society *so that having a major impairment such as deafness imposes no limitations* on individuals' opportunities are not so easily dismissed. Those costs count from a moral point of view, because there is a morally legitimate interest in avoiding them [i.e. avoiding the costs]. (FCC, pp. 283–284, emphases added)

The last line of this argument asserts that there is a morally legitimate interest in avoiding the costs of a society in which all impairments were accommodated to such a point that they produced absolutely no disadvantages to people who possessed them. I will not dispute this claim. However, I will dispute the claim that the DR movement demands such remedies, and that therefore the demand for such remedies reduces the moral legitimacy of the movement. The argument misses the point of the DR demand for justice.

Recall the distinction between CDIs (socially mediated disadvantages) and UDIs (unconditional or innate disadvantages). Every argument that DR advocates have ever made distinguishes between these two categories of disadvantage. For example, the oldest distinction in the movement is embedded in the distinction between "impairments" (defined as brute biomedical facts about individuals) and "disabilities" (then defined as the disadvantages caused by social arrangements to people who have impairments; see UPIAS, 1976). The movement is *only* concerned to remove "disabilities"—that is, CDIs. Impairments are assumed to cause other disadvantages (that I have labeled UDIs), but the movement simply doesn't discuss those. It doesn't even have a label for them—which is why I had to invent the clumsy term UDI! (*Impairment* is a label for the biomedical condition, not for the disadvantages that impairments inherently entail—if any.)

FCC's argument is that the DR movement differs from the movement for blacks and gays *because* "disadvantages remain" after discrimination is removed, and that it would be unjust to impose the costs of removal of *these* disadvantages (UDIs, the ones that remain, the disadvantages that are inherent to impairments) on society. But wait—the DR movement has never asked that UDIs be removed! The movement is *only* interested in CDIs— "disabilities" as they were designated in the 1976 UPIAS definition. The disadvantages that come from something *other than* discrimination are of little interest to the movement, and are surely not a basis for a justice claim of remediation. The general attitude of the movement is that impairment, in and of itself, is something that we can live with. (And why do nondisabled people make such a big deal out of it, after all?) The demands of the movement are not to remedy all disadvantages, but merely to remedy those that are discriminatory, caused by society, and therefore are society's responsibility. When the society builds sidewalks that cannot be used by people in wheelchairs, that's discriminatory. But the fact that Mount Whitney is inaccessible to wheelchair users is not discriminatory, and no one has ever (contrary to the politically-exciting nightmares of conservatives) claimed that the mountain should be ramped.

As already acknowledged, the demarcation between CDIs and UDIs is not cut and dried. But drawing the line between *what is* and *what is not* society's responsibility is a problem for all civil rights movements, not only for DR. These debates are ongoing. Affirmative action is one example, of interest to racial and ethnic minorities. The question of social responsibility for day care of children is another, of interest to the women's movement. The DR movement will presumably have to engage in similar contests with the status quo. But the mere fact that some impairments involve UDIs makes absolutely no difference to the legitimacy of DR as a civil rights movement, contrary to the quoted argument from FCC. If the authors had better understood the nature of the movement, and not been distracted by the scary but irrelevant facts of UDIs, they could not have made such an argument.

10.4 Actual Arguments 2: "Not Unduly Burdensome"

Case 5. Our society has learned through its efforts to accommodate people with disabilities that in many cases lowering the barriers to participation *need not be unduly burdensome* to others. (FCC, p. 320, emphasis added)

Case 5 gives the appearance of an endorsement of the goals of the DR movement. By now the reader knows what to expect: I will challenge this statement's dedication to the goals of the DR movement. Just so. I will compare the statement to similar statements as they would apply to civil rights for racial minorities and women.

One problem in comparing this statement to similar statements regarding sex or race is that the talk about "lowering barriers" only suits a few cases, such as lowering the employment requirements for carrying weight for women firefighters. Most cases of integration are different: integrating a lunch counter is not a matter of *lowering* a barrier, but of removing it. The expression "lowering barriers" makes it sound as if the barriers were there *for a purpose*, so that lowering them is itself some sort of a compromise with high performance. This would be the case if the barriers were in the nature of high standards, like weight-carrying requirements for firefighters, or a grade point average required for entry to a college. But, according to DR activists, the barriers to the participation of people with impairments are not high standards at all. Instead they are arbitrary barriers and obstructions in the environment that serve no legitimate purpose at all. Being able to climb a set of stairs is a requirement for the job of firefighter, but not for the job of receptionist. So *removing* the barrier of a stairway entrance would be a suitable accommodation for the job of receptionist. (Under the Americans with Disabilities Act, the removal of stairway barriers is subject to certain cost considerations, which will be discussed below.) Sidewalks without curb cuts, and television programs without captioning, are arbitrary barriers that need to be *removed*, not lowered. They are not high standards that encourage high performance and yield high social benefits. No public interest is ensured by keeping paraplegic people at home and deaf people uninformed. So the phrase "lowering of barriers" already encourages a misleading notion of barriers in the context of DR. So I will remove the talk about barriers. Let us compare the expressed views about disability with parallel views about women's rights and rights of minorities.

Case 5a. Our society has learned through its efforts to integrate people with disabilities into the workforce that in many cases doing so need not be unduly burdensome to others.

Case 6. Our society has learned through its efforts to integrate women into the workforce that in many cases doing so need not be unduly burdensome to men.

Case 7. Our society has learned through its efforts to integrate African Americans into the workforce that in many cases doing so need not be unduly burdensome to white Americans.

Case 5a is exactly parallel to Cases 6 and 7. But something sounds very wrong in Cases 6 and 7. What is it?

For a start, are we willing to say that the integration of women and nonwhite races is justified only "in many cases"? No. To affirm civil rights integration only "in many cases" is to suggest that in many (perhaps most) other cases, integration of women and minorities is outweighed by the "burden to men" and the "burden to white Americans." No civil rights advocate would claim this. Nevertheless, FCC makes exactly that claim about disability. This is an extraordinarily grudging acknowledgement that disabled and nondisabled people might (sometimes, somewhere, maybe) live in the same integrated world.

Finally, and most importantly, notice that Case 6 weighs the integration of women against the burdens *on men*; Case 7 weighs the integration of African Americans against the burdens *on white Americans*. Is this how we think about integration? I say no. It is divisive. Civil rights (at least as seen by civil rights advocates) does not pit the interests of one group against another group, women against men and blacks against whites. But Case 5 (and 5a)—an ostensibly pro-DR statement—pits the interests of people with impairments against the burdens experienced by "others," that is, by nondisabled people! It's US against THEM—nondisabled people against disabled people. The only way disabled people can justly expect integration (according to the authors of Case 5) is if integration is not unduly burdensome *to nondisabled people*!

These authors are speaking about disability in a way that they would never speak about other discrimination. No real civil rights advocate would divide the interests of minority and majority groups in this way. Just as Cases 6 and 7 would not be made by someone who was a genuine advocate of civil rights for women and minority races, Case 5 would not have been made by a genuine advocate of DR. It is a condescending and divisive statement by a nondisabled person, acknowledging only that *sometimes* the rights of *those* disabled people do not harm *us*. Only in the case of disability rights is the academy so backwards in its thinking.

10.5 Actual Arguments 3: The Meaning of "Reasonable Accommodation"

Case 4b. The fact that it is costly to remove barriers of discrimination against blacks or gays has no moral weight because no one can have a morally legitimate interest in preserving unjust arrangements. (Achieving a fair distribution of the costs of reform is another matter, of course.) (FCC, p. 284)

Case 4b is a modified quotation of the passage quoted in Case 4a. In the earlier instance the parenthetical sentence was replaced with ellipses because it was irrelevant. In the present context it is crucial. It expresses a reservation regarding the costs of removing unjust social barriers for blacks and gays. It is said that, even though no one can have a morally legitimate interest in preserving unjust barriers, under

certain circumstances unjust barriers might justly remain in force. These circumstances have to do with the costs of reform, and finding a just method of distributing
these costs. The outcome may be that unjust social barriers remain in the society
because no means can be found of *justly* allocating the costs of removal. An example
might be the (alleged) injustice of job quotas. Quotas would more efficiently remove
the injustice of unequal employment among races and sexes better than would the
present (nearly unenforceable) prohibitions against bias in hiring. But quotas would
do so at the cost of (allegedly) unjust decisions to favor an individual for a job on
the basis of the person's membership in a disadvantaged group. The just results, of
equal employment opportunity, is (at least) delayed because accelerations of the process would require unjust hiring practices. A second, simpler example would be the
fact that large corporations have broader legal responsibilities to document equal
opportunity hiring practices than small companies. Even though discrimination is
equally unjust in large and small companies, the costs of assuring nondiscrimination
are (allegedly) more justly borne by large companies than small companies. These
observations are reasonable. There is room for disagreement about what counts as a
"just distribution" of the costs of reform, of course.[4] But the principle is correct; the
reform of injustices must be designed not to create too many new injustices, and if
such reforms cannot be devised, then the old injustices may "justly" remain in place
longer than they would if a just reform were possible.

Now we are ready to consider how FCC interprets the concept of "reasonable
accommodation" in the context of the Americans with Disabilities Act (ADA).

Case 8. It is important to emphasize that the ADA adds the qualifier that all that is required
in the name of equal opportunity is 'reasonable' accommodations. The addition of this
qualifier signals a recognition that the interest of employers, of workers who do not have
disabilities, and of consumers of the goods and services that public and private organizations
produce are also legitimate and should be accorded some weight. (FCC, p. 292)

Notice the contrast between Case 4b and Case 8. Case 4b says that *no one can
have a legitimate interest* in preserving unjust arrangements against blacks and gays
(although the difficulties of achieving a fair allocation of costs may complicate the
removal of unjust arrangements). But when we discuss people with impairments,
Case 8 claims that employers and nondisabled workers *do have a legitimate interest*
in maintaining arrangements that segregate disabled people. Evidence in favor of
this legitimate interest is said to be found in the ADA's reference to "reasonable
accommodations," a term which limits the immediate responsibilities of employers
and places of public accommodation to provide integration. This interpretation of
the term "reasonable accommodations" is one of several arguments one can find in
FCC that disability rights are not on a par with the rights of women and minorities.

Grassroots disability rights workers argue on the street (literally on the street,
because the buildings at issue are inaccessible) with restaurant owners and employers in an attempt to gain access to the goods of our society. Such workers hear
the term "reasonable accommodation" incessantly. "Reasonable" is always given
heavy emphasis. People who manage inaccessible facilities seem to believe that the

term "reasonable" is free license within the law to give no accommodation at all—after all, isn't the business owner the best person to decide what is *reasonable*? Isn't delivering a meal in a paper bag to a wheelchair user in front of the restaurant a "reasonable" substitute for installing a ramp (that may cost $500) to allow the person to actually sit in the restaurant in the company of nondisabled people, perhaps his or her friends? According to FCC, the restaurant owner and nondisabled patrons have a legitimate moral interest in opposing the wheelchair user's right to equal access, even though they do not have a legitimate moral interest in opposing the rights of racial and other minorities to equal access. Disability rights are second-class rights.

But what other explanation could there be of the term "reasonable accommodation" in the law? What purpose could the term "reasonable" serve other than acknowledging the legitimacy of the opponents of equal access for people with impairments? FCC has already given a perfectly good answer to this question. The reference to *reasonable* accommodation need not mark the moral legitimacy of those who oppose equal access. Instead, it marks the difficulties in achieving a fair distribution of the *costs* of reform. This is clearly stated in Case 4b, as applied to the costs of removing the unjust barriers encountered by blacks and gays. When blacks and gays are forced to wait long periods for justice, the authors say that *no one can have a morally legitimate interest in preserving those unjust arrangements* (although the difficulty of justly allocating costs can lead to delayed justice). But when the same thing happens to people with impairments, the authors say that *employers and nondisabled people have morally legitimate interests in preserving unjust arrangements* for disabled people. Why should the same principle that is applied to women and minorities not be applied to people with impairments?

This is an obvious double standard. The authors have a perfectly good analysis of why injustices are sometimes not immediately resolvable—an analysis that *does not* imply that majorities have a legitimate interest in denying the rights of minorities. Prohibitions on hiring quotas for minorities may slow down justice, but they do not *deny* justice (because no one can have a legitimate interest in preserving unjust arrangements). The very same rationale could have been given for "reasonable accommodations" for disability. But the authors chose not to offer the same protection to disabled people that they offer to other civil rights groups. Instead, they claimed that the expression "reasonable accommodations" indicates that employers *do have* a legitimate interest in preserving the unjust segregation of disabled people, even though the same employers *do not have* a legitimate interest in preserving the unjust segregation of blacks, gays, and women. The DR movement does not share the legitimacy of other civil rights movements. There are many claims for equal access to the goods of society, but some are more equal than others.

10.6 Conclusion

My intent is to illustrate how the DR movement has less support within academic discourse than civil rights for women or other disadvantaged groups. I submit that John Cleese would not have performed or endorsed a Zip Coon-era joke that exploited racist assertions about verbal incompetence, but he did perform and endorse an identical joke about people with cerebral palsy. I submit that the authors of FCC would not have claimed that the integration of women and racial minorities is merely "in many cases not unduly burdensome to others," but they make exactly that claim about disabled people. Nor would these same authors have claimed that dominant groups (men or racial majorities) have a morally legitimate interest in maintaining segregationist arrangements against women or racial/ethnic minorities. But they claim that nondisabled people have exactly that interest in maintaining the segregation of disabled people.

The academy may not genuinely accept equality for women, racial/ethnic minorities, or gays and lesbians. However, the discourse of academia does, at least, pretend to respect those civil rights. It does not even pretend to respect similar rights for people with impairments. If other civil rights groups are currently at the level of *merely verbal* support from the academy, the DR movement is at the level of *not even verbal* support for its rights.

Notes

1. An example is the "storm" of protest in response to former Harvard President Lawrence H. Summers's comments in 2005 suggesting that the low number of female science faculty might be due to innate differences between the sexes concerning science and math abilities (Dillon & Rimer, 2005).
2. This terminology follows Amundson and Tresky (2007). It will be noticed that this is a version of the impairment/disability distinction used within the DR movement. This version avoids the terminological confusions caused by the convention that "disability" *means* socially-conditioned disadvantages of impairments.
3. Indeed, I used the same oppressive technique as Fanon in an early DR paper, stigmatizing others as a means of defending oneself against stigma. I argued that people with impairments should not be treated like ill people or frail elderly people. See Amundson (1992) and Wendell (1996, Chapter 1, as a corrective).
4. In fact I do not agree with these analyses; quotas are perfectly fine with me. But they appear to be the kind of thing that FCC is alluding to in distinguishing between justice and the costs of remedying injustice.

Acknowledgements Research was supported by National Institutes of Health grants No. S06-GM08073 and R03-HG3632-01A1. I am also indebted to Shari Tresky for research and conceptual development.

References

Amundson, R. (1992). Disability, handicap, and the environment. *Journal of Social Philosophy, 23*, 105–118.
Amundson, R., & Tresky, S. (2007). On a bioethical challenge to disability rights. *Journal of Medicine and Philosophy, 32*, 541–561.
Buchanan, A. E., Brock, D. W., Daniels, N., & Wikler, D. (2000). *From chance to choice: Genetics and justice*. Cambridge, UK: Cambridge University Press.

Canadian Association of Professionals with Disabilities. (2007). Web page (accessed March 29, 2007). Available at http://www.canadianprofessionals.org

CBS Morning Show. (2001). *Interview with Jerry Lewis*. Broadcast May 20, 2001.

Cleese, J. (2006). *Profile: John Cleese* (Interviewer M. Bragg). Ovation Channel. Broadcast December 18, 2006, copyright 1986.

Darke, P. (1998). Understanding cinematic representations of disability. In T. Shakespeare (Ed.), *The disability reader: Social science perspectives* (pp. 181–197). London and New York: Continuum (Cassell Educational Ltd.).

Dillon, S., & Rimer, S. (2005). No break in the storm over Harvard President's words. *New York Times*, January 19, 2005. Available at http://www.nytimes.com/2005/01/19/education/ 19harvard.html

Dormon, J. (1988). Shaping the popular image of post-Reconstruction American Blacks: The "coon song" phenomenon of the Gilded Age. *American Quarterly, 40*, 450–471.

Fanon, F. (2001). The lived experience of the black. In R. Bernasconi (Ed.), *Race* (pp.184–202). Malden, MA: Blackwell Publishers. Translated by V. Moulard from F. Famon, L'expérience vécu nuNoir. Esprit, 19(179), May 1951, pp. 657–679 by permission of Esprit.

Field, S. (1994). *Screenplay: The foundations of screenwriting*. New York: Dell Publishing.

Goffman, E. (1963). *Stigma: Notes on the management of spoiled identity*. Upper Saddle River, NJ: Prentice-Hall.

Henderson, H. (2006, November 11). Why the gap between ability, job quality?' Toronto Star, sec. Life, p. L.3.

Longmore, P. K. (1985). Screening stereotypes: images of disabled people. *Social Policy, 16*, 31–37.

Norden, M. (1994). *The cinema of isolation: A history of physical disability in the movies*. New Brunswick, NJ: Rutgers University Press.

Riggs, M. (1986). *Ethnic notions* (video). Marlon Riggs (Director). California Newsreels Publishers. Transcript at http://newsreel.org/transcripts/ethnicno.htm

Sontag, S. (1989). *Illness as metaphor & AIDS and its metaphors*. New York: Anchor.

UPIAS (1976). *Fundamental principles of disability*. London: Union of the Physically Impaired Against Segregation. Reprinted in M. Oliver (1996). *Understanding disability: From theory to practice*. Houndmills, UK: Palgrave.

Wendell, S. (1996). *The rejected body: Feminist philosophical reflections on disability*. New York: Routledge.

Chapter 11
Dignity, Disability, Difference, and Rights

Daniel P. Sulmasy

In recent decades, disabled persons living in Western nations have joined forces to fashion a powerful, organized, political force, pressing the case for their right to equality of opportunity in society. As Shapiro (1994, p. 11) puts it, "...alongside the civil rights struggles of African-Americans, women, gays and lesbians, and other minorities, another movement has slowly taken shape to demand for disabled people the same fundamental rights that have already been granted to other Americans." Although challenges remain, this movement has been highly successful. Concrete achievements in the U.S. have included the passage of the Americans with Disabilities Act, cuts in curbs at intersections, kneeling buses, and wheelchair accessible toilets. The political and legal methods employed by the disability rights movement have emulated those of earlier civil rights movements. These political and legal techniques were "learned from the civil rights movement and the nascent women's movement." (Shapiro, 1994, p. 47).

The basic structure of the argument employed by all of these political and legal movements can be outlined according to a common formula. Let us call it the Standard Civil Rights Formula. The structure of the argument may be represented as follows:

1. Z-people are not respected, treated as equals, or granted full rights.
2. Treating Z-people differently from the general population is commonly justified by arguing that having the Z characteristic makes Z-people unequal.
3. But the Z characteristic does not make people *unequal*, just different.
4. Prejudice may be defined as the unequal treatment of people based upon a particular characteristic that makes them merely different, such as Z.
5. Prejudice is morally wrong.
6. Therefore, a good society, free of prejudice, considers Z-people merely different, but not unequal, and affords Z-people equal respect and rights.

D.P. Sulmasy (✉)
MacLean Center for Clinical Medical Ethics, University of Chicago, Chicago, Illinois, USA
e-mail: dsulmasy@medicine.bsd.uchicago.edu

D.C. Ralston, J. Ho (eds.), *Philosophical Reflections on Disability*, Philosophy and Medicine 104, DOI 10.1007/978-90-481-2477-0_11,
© Springer Science+Business Media B.V. 2010

This logic has been rehearsed in many settings. Arguments using the Standard Civil Rights Formula have been advanced, for example, when Z has been instantiated as black, Latino, female, gay, or disabled, just to name a few. On the strength of this formula, laws have been changed and previous victims of prejudice have made great strides in gaining public acceptance, approbation for their political demands, socioeconomic progress, and legal rights.

For the disabled, this form of argument has been highly effective. Sadly, those who are disabled frequently have been labeled "defective," and this designation commonly has been interpreted as an inequality justifying unequal treatment. Historically, the idea that the disabled are physically unequal has been used to justify unequal social structures and practices that have inhibited disabled persons from flourishing as best they can as human beings who are recognized, in some deep sense, as equals. In the past few decades, however, having forcefully made the argument that they are *not* unequal because of their disabilities, but merely different, and that differences need to be respected in a just society, the disabled have made great strides towards social justice. The Standard Civil Rights Formula seems to have worked well for them.

Prejudice is deeply immoral and the elimination of prejudice is a very good thing for a society. Nonetheless, one might ask whether this Standard Civil Rights Formula is really sound. While acknowledging that it has been used to accomplish much that is good, one might ask whether it has not also led to certain practices that are, in fact, harmful to the common good. For instance, at least to some, the substitution of a neologism such as "differently-abled" for "disabled," a linguistic shift that is supported by the logic of Standard Civil Rights Formula, suggests a denial of reality. That seems a very high price to pay in exchange for greater recognition of dignity and rights. In addition, in the view of many, certain more recent arguments that seem to follow from the Standard Civil Rights Formula seem to constitute sufficient reasons for doubting the soundness of this formula. For instance, having come to regard deafness as a difference to be respected rather than a defect to be avoided, parents have employed genetic techniques to selectively bear and rear children who are deaf. The permissibility of this practice follows from the logic of the Standard Civil Rights Formula. Yet many find this practice at least as immoral as prejudice against the deaf, leading them to wonder whether the Standard Civil Rights Formula might be a house built on sand.

It is, of course, a fact that human beings are different. Human beings all differ genetically. Even identical twins have small differences in the imprinting of genes, biochemical modifications that transpire over development. Human environments differ—physically, biologically, familially, socially, economically, religiously, educationally, and culturally—and these differences also contribute to human diversity. Subsequently, human beings differ in their appearance, physical capability, health, personality, intellect, judgment, emotional reserve, financial well-being, etc. How can one make a claim for equality in the face of so many obvious differences?

One way to establish equality might be to make moral claims on the basis of something that all human beings have in common. The Standard Civil Rights Formula does not pursue this tack. The Standard Civil Rights Formula is agnostic about whether human beings have anything in common other than difference.

A philosophical climate hostile to essentialism has made it virtually impossible to say anything else. This has left few options. It has seemed, instead, that the only way forward has been to assert that the obvious differences one encounters between human beings do not mark a moral difference—that the alleged fact of inequality is not a basis for inequality in treatment. This is the approach of the Standard Civil Rights Formula.

Yet, upon probing a bit deeper, it seems as if this approach is difficult to defend. There are only a few limited options available to those who wish to justify the claim that no differences between human beings mark any moral differences.

One strategy might be to axiomatize the claim that the fact of inequality is not a basis for inequality in treatment. But on the face of it this seems to be a very odd claim—one that violates the ancient Aristotelian principle of formal justice that similars should be treated similarly and dissimilars according to their relevant dissimilarities. The claim would thus seem far from intuitively obvious—a claim in need of further justification and not an axiom.

Another approach might be to seek justification of the claim by resort to a prior principle of justification by autonomous assertion—suggesting that the underlying justification is the principle that any difference an individual does not believe sufficient to justify differential treatment is not a legitimate moral or legal basis for differential treatment of that individual. While it may at times appear that some contemporary arguments can be construed this way, such a purely subjective account would be a formula for chaos and cannot be correct. Imagine that characteristic Z is having homicidal thoughts and plans. Plainly, such a difference merits differential treatment. The mere subjective assertion that one's difference (Z) does not justify differential treatment is insufficient. One cannot appeal to pure moral subjectivism.

Of course, a more careful account would limit subjective appeals by restricting the differential treatment to only those differences with the potential to harm others. But this, in turn, leads to several conundrums for those who would appeal to this principle to justify a robust set of rights for the disabled. This sort of Millian appeal to liberty, limited only by harm to others, really only justifies *negative* rights. If characteristic Z is quadriplegia, then this appeal would certainly justify a prohibition on dumping quadriplegics in the river, but it is hard to see how it could justify a social duty to provide them with motorized wheelchairs. Further, harm could be construed broadly by other members of society. The time, expense, and energy needed to care for the disabled takes resources away from those who are not disabled, and this could be construed as a harm to the majority. This approach could therefore lead to a diminution in rights for the disabled.

Another strategy might be to seek intersubjective social agreement on which different characteristics merit differential treatment and which do not. This amounts to a full-scale retreat from the principle that differences do not mark a moral warrant for differential treatment. Yet, the hope that society will intersubjectively agree to a narrow range of differences justifying differential treatment has made this the most common strategy employed in current debates.

The commonly available strategies for obtaining intersubjective agreement about which different human characteristics justify differential treatment, however, are not based on any universally accepted moral principles and certainly do not assure a robust set of rights for the disabled. For example, Utilitarianism offers a solution for deciding which differences should be afforded rights and which should not. Since the basis of Utilitarianism is the maximization of the net social utility, however, the disabled typically fare quite poorly in Utilitarian arguments. The inability of Utilitarianism to protect what most persons intuitively sense to be the just claims of minorities (such as the disabled) is often cited as one of the major problems with Utilitarianism as a system of morality. As Rawls (1971, p. 27) once pithily described the problems Utilitarianism has in accounting for justice, "Utilitarianism does not take seriously the distinction between persons." Disabled persons, and many others concerned with a robust account of justice, are not likely to see Utilitarianism as the solution to the problem of how to decide which differences should be respected and which should justify differential treatment.

Alternatively, one might take a deconstructionist approach and argue that all there is just is difference, and that what presents itself as an argument for differential treatment based upon justice is really only power and domination masquerading as justice. By "unmasking" the hypocrisy of those who make such claims and by making counter-assertions of "disabled power," the disabled can wrest rights from the entrenched class of the powerful and privileged. Disabled persons have often employed such arguments successfully to justify the application of the Standard Civil Rights Formula to their struggle for rights. Disabled persons, however, should be very wary of such approaches. While space limitations do not allow a full explanation here, deconstruction is, in the view of many, an unsound philosophical system. If deconstruction is itself "unmasked," a deconstructionist argument for rights for the disabled collapses. Further, the conditions under which disabled persons are able to assert "disabled power," for instance, by protest, or chaining themselves to doors of the U.S. Supreme Court, or blocking access to public transportation, etc., are ephemeral and tenuous. Only in certain kinds of very wealthy and liberal states can such a strategy succeed. Political and economic conditions can change rapidly, however, and the "power" of the disabled can be quashed easily, leaving the disabled with no philosophical basis on which to argue for their rights. In the end, the *argumentum ad baculum* remains an informal logical fallacy and an incredibly weak defense for any system of human rights.

Thus, it seems reasonable to see if there are other, sounder ways to understand disability and difference that might justify a robust account of rights for the disabled without entailing strange, if not absurd, consequences.

11.1 Natural Kinds

A new concept has begun to emerge in contemporary analytic philosophy that provides the foundation for an alternative view of disability, dignity, and rights for the disabled. This is the notion of natural kinds.

Kripke's work on identity and necessity (1971, 1972) is credited as the starting point for contemporary discourse about natural kinds. Kripke has argued that "identity statements are necessary and not contingent." He has used the term "rigid designator" to describe a term that might previously have referred to an essential feature. A rigid designator is "a term that designates the same object in all possible worlds." Kripke does not mean by this that the thing must *necessarily* exist in all possible worlds, but rather that if it were to exist in any possible world, the rigid designator would designate it in that world.

Baruch Brody's analysis (1980, pp. 100–134) is more forthrightly essentialist and Aristotelian. Brody (1980, p. 111) argues that "identity across possible worlds is prior to rigid designation," and that something must already have been picked out as the same in any possible world before it can be designated as such. Applying this to the naming of the kinds of things there are in the world, this means that there is something about each natural kind that is common to all members of the kind and yet distinct from all other kinds—a "kindedness," if you will—that precedes the naming of the kind. Language and reference have a relationship, and in the case of natural kinds, that relationship is fixed by reality. Correctly describing the natural world requires a mind-to-world "direction of fit"—i.e., the mind must conform itself to the world rather than the world being shaped by ideas in the mind.

Natural kind thinking is not foreign to bioethics. For example, the notion of substituted judgment requires that we assume that the name "Mr. Smith," designating a man who is comatose, also rigidly designates the Mr. Smith to whom we refer in that possible world in which comatose patients can tell us their preferences. Or, to take another example, as Laura Garcia (2008) notes, even to speak of "animal rights" presumes a distinction between the human natural kind and other animal kinds.

Kripke's work has been extended by David Wiggins (1980, 2001). Wiggins has developed the notion of "sortal predicates" by which entities of a certain kind are picked out, identified, and re-identified over time. He has come to the conclusion that the predicate calculus simply cannot account for much of *what is* (particularly living things) unless it is enriched by the addition of the concept of a sortal predicate (1980, p. 38). In other words, one cannot say "Smith is the same man I saw yesterday" without predicating that Smith and the man I saw yesterday belong to the same natural kind (in this case, human beings). Thus, Wiggins says, we must embrace at least a "modest" form of essentialism (2001, pp. 107–138). These essences are not Platonic forms. As Wiggins puts it, essences are not "fancied vacuities parading in the shadow of familiar things as the ultimate explanation of everything that happens in the world. They are the natures whose possession by their owners is the precondition of their owners being divided from the rest of reality as anything at all." (1980, pp. 132–33).

Wiggins says very clearly, if densely,

> All the doctrine implies is that the determination of a natural kind stands or falls with the existence of law-like principles, known or unknown, that will collect together the extension of the kind around two or three good representatives of the kind (2001, p. 80).

There are law-like principles that collect together the actual extension of those individuals one calls human beings. These law-like principles "determine, directly or indirectly, the characteristic development, the typical history, the limits of any possible development or history, and the characteristic mode of activity of anything that instantiates the kind" (2001, p. 84). Amazingly enough, the average man or woman (or even child) is easily able to tell which entities belong to this natural kind, and which do not, in the absence of technical biological knowledge. One can also readily recognize those deviations from the characteristic development and typical history that render some members of this natural kind defective. So, children born with phocomelia (foreshortened limbs) are defective members of the human natural kind. They are not members of some other natural kind, say walruses. Without at least this much essentialism, medicine would not even be conceptually possible.

11.2 Diseases and Natural Kinds

The notions of disease and injury are important in considering disability. Most disabilities result from disease or injury. I have argued elsewhere (Sulmasy, 2005) that a disease is:

> A class of states of affairs of individual members of a living natural kind Y, that:
> (1) disturbs the internal biological relations (law-like principles) that determine the characteristic development and typical history of members of the kind, Y,
> (2) in a pattern of disturbance shared with at least one other member of the kind, Y,
> (3) The aim of this classification must be to provide at least a provisional basis for explaining the causes and/or natural history of a disturbance in the internal biological relations of the affected members of Y, and
> (4) at least some individuals of whom (or of which) this class of states of affairs can be predicated are, by virtue of that state, inhibited from flourishing as Ys.

This definition can be extended to cover injuries as well (Sulmasy, 2005).

One should take note of some of the subtleties in this definition. A disease is not a natural kind. It is a classification of a state of affairs that can occur in members of a particular living natural kind. Further, the "biological internal relations (law-like principles)" to which the definition refers encompass anatomy, physiology, and psychology. Moreover, the primary purpose of disease classification (nosology) is explanatory. Even if the disease does not provide a causal explanation for the illness, the purpose of bringing a pattern of disturbance under a particular name is to predict an expected natural history and provide the first step towards pathological explanation. One often hopes that this explanatory knowledge will lead to better treatments. But as a scientific concept, the primary purpose is explanatory.

Finally, one must especially note that setting the *telos* as "the flourishing of the individual as the kind of thing that it is" also explains why it can be controversial to classify as diseases certain patterns of variation in the law-like biological principles that determine the characteristic development and typical history of a living natural kind. In particular, patterns of human behavior are most susceptible to being controversially called diseases. But this does not undermine the definition of a disease. It is only to say that the task of deciding whether to use the word "disease" to designate a pattern of variation in the law-like principles that govern a thing as a member of a kind will have some very clear cases and some not so clear cases. For example, lung cancer is almost universally thought to interfere with human flourishing. It is uncontroversially accepted as a disease. Other cases are much more contestable. For example, homosexuality would not universally be thought to interfere with human flourishing.

The fact that there are such contested cases does not imply that the definition of disease is a "subjective value judgment." For the human natural kind, such arguments are anthropological, in the philosophical sense of this word. That is to say, these are arguments over what kind of a thing a human being is and what constitutes flourishing of the individual as a human. The fact that there is uncertainty about how one can tell whether a human being is flourishing or that there are arguments in philosophical anthropology over what sort of thing a human being is do not imply that what constitutes flourishing as a human being depends solely on human choice and has nothing to do with what a human being is.

11.2.1 Realism and Anti-Realism

Diseases are not primary existents. "Down syndrome" does not pick out a primary existent, but a class of states of affairs occurring in members of a natural kind. Diseases are not natural kinds, but states of affairs. Diseases have no essences.

Saying this does not imply, however, that diseases have no objective basis, or are merely human constructions, or that diseases are merely value judgments. Diseases make necessary reference to natural kinds, and natural kinds admit of at least a modest essentialism. It is essentialism about living natural kinds and their natural dispositions that provides the foundation for realism about diseases. The patterns of disturbance that one classifies as diseases are not arbitrary. There is a mind-to-world direction of fit. It is the pattern of disturbed internal biological relationships in the natural kind that imposes itself upon the observer; the observer does not impose the pattern upon the affected members of the kind. The world is the standard by which the observer's beliefs are judged.

11.3 Disability, Disease, and Natural Kinds

Suppose that it is a law-like generalization and typical feature of natural kind Y that its members have capability C, and that C regularly enables members of Y to accomplish q. The main way that diseases and injuries inhibit flourishing is by diminishing capabilities such as C. Accordingly, consider two individual members of the natural kind Y, X_1 and X_2. Suppose that X_2 suffers from a disease or injury that interferes with capability C. Disability can then be described formally as follows:

> If, given standard conditions (cf. Nordenfelt, 1995, pp. 65–80), the effort of X_1, by virtue of capability C, to accomplish q is much, much less than the effort of X_2, then X_2 is disabled with respect to q.

For example, among human beings, suppose that q is the ability to understand spoken language, and that C is the capability to hear. Suppose that X_2 suffers from the disease of congenital bilateral cochlear nerve deficiency and therefore lacks the capability of hearing (C). Since, given standard conditions, it would only be with great effort that X_2 could learn to read lips and understand spoken conversation compared with X_1 who has no disease or injury affecting hearing, then X_2 is disabled with respect to the ability to understand spoken language.

In a technical sense, even the common cold "disables" a person, at least for a short time. Typically, however, one reserves the adjective "disabled" for those whose disability persists indefinitely and for whom the capability (C) and goal(s) q are significant.[1]

It is also important to point out that, necessarily, it is by noticing the deviation from the law-like generalizations and typical features and history of members of the kind that one makes the diagnosis of a disease (congenital cochlear never deficiency) in X_2, and concludes that X_2 is disabled. Further, as I will argue below, it is by virtue of the fact that X_1 and X_2 are members of the same natural kind that one can conclude that both have rights and that X_2 is owed treatment (if possible) and social accommodation for that disability.

11.4 Natural Kinds and Dignity

The word "dignity" is used in many different ways, but I argue that the notion of natural kinds is also central to the idea of dignity. Elsewhere (Sulmasy, 2008), I have designated three different ways of talking about dignity as the *attributed*, the *intrinsic*, and the *inflorescent* meanings. *Attributed* dignity refers to the way in which we use the word to confer a special value on individuals or states of affairs. To refer to someone as a visiting dignitary, or to say that a particular task is undignified, is to use this word to refer to an attribution, based upon individual or collective choices, of a certain level of value to an individual. By contrast, *intrinsic* dignity makes necessary reference to a particular natural kind, that, as a kind, is possessed of a certain high degree of value that we designate by the word "dignity." This intrinsic meaning of dignity refers to the value human beings have by virtue of being the kinds of things that they are—human beings. This is the meaning of dignity invoked, for instance, by the opening sentence of the United Nations Universal Declaration on Human Rights (1948), which states that, "recognition of the inherent dignity and of the equal and inalienable rights of all members of the human family is the foundation of freedom, justice and peace in the world." The word, here, is being used to refer to the *intrinsic* meaning of dignity. The *inflorescent* meaning of dignity refers to those uses of the word that evoke a sense of human excellence or virtue. When one says, "He faced that situation with great dignity," or, "She is a truly dignified and gracious human being," one is referring to an inflorescent meaning of dignity.

These three uses of the word "dignity" are by no means mutually exclusive. However, the primary moral meaning of dignity is the intrinsic meaning. One would not refer to either the inflorescent or the attributed dignity of an individual member of a natural kind unless that kind were possessed of intrinsic dignity. For instance, one does not, in an inflorescent sense, describe the pattern of flight of a mosquito as dignified. Nor does one say, in an attributed sense, that being zapped by an electronic insect control device is beneath the dignity of a mosquito.

What are some of the features that lead one to conclude that the intrinsic value of a natural kind is sufficient to warrant the appellation of dignity? First one need only accept the commonly espoused view that there is a hierarchy of natural kinds. This need be no more than an acknowledgment that there are some forms of life that are "higher" than others—for example, comparing an ameba with a human being. One then only need take note that at least one natural kind has kind-typical capacities for language, rationality, love, free will, moral agency, creativity, humor, and an ability to grasp the finite and the infinite. Ordinary use of the word suggests that members of natural kinds that have kind-typical capacities for language, rationality, love, free will, moral agency, creativity, humor, and an ability to grasp the finite and the infinite are worthy of this appellation. This is not a "speciesist" account. If there are angels or extraterrestrial natural kinds with these capacities, they also would have dignity. But it is at least clear that these are refined capacities and that they typify the human natural kind, and not mosquitoes.

The bottom line is this: intrinsic dignity, the fundamental moral worth or value of a human being, is based upon nothing other than the bare fact that one is a member of a natural kind, that, as a kind, is possessed of these features. As such, intrinsic dignity is absolutely equal, inalienable, and does not admit of degrees.

11.4.1 Intrinsic Dignity and Rights

As I have noted, the concept of intrinsic dignity implies the notion of a natural kind. This link between intrinsic dignity and membership in the human natural kind, in turn, builds an argument for dignity, rights, and equality based upon a completely different foundation than the Standard Civil Rights Formula. If intrinsic dignity is the value an individual has by virtue of being the kind of thing that it is, then that value is not based upon any characteristic "Z," but upon the bare fact of membership in the natural kind. In other words, it is not being black or white, male or female, able or disabled that grounds arguments for dignity. Rather, all that counts is being human.

Importantly, the logic of natural kinds suggests that one picks out individuals as members of the kind not because they express all the necessary and sufficient predicates to be classified as a member of the species, but by virtue of their inclusion under the extension of the natural kind that, *as a kind*, has those capacities. In technical language, this is extensional, not intensional, logic. For example, very few lemons in the bin in the supermarket express all the necessary and sufficient conditions for being classified as fruits of the species *Citrus limon*. We define a lemon as a yellow fruit. Yet some specimens in the bin are yellow, some are green, some are spotted, and some are brown. Nonetheless, they are all lemons.

The care of the disabled depends profoundly upon this extensional logic. For instance, one might, say, define the species *Homo sapiens* intensionally, as a rational biped. However, it is not the expression of rationality or the ability to walk on two legs that makes an individual human, but that individual's belonging (extensionally) to a kind that is, *as a kind*, characterized by its capacities for rationality and bipedal ambulation. When a human being is mentally retarded or paraplegic, we first pick the individual out as a human being, then we note the disparity between the characteristics of the afflicted individual and the paradigmatic features and typical development and history of members of the human natural kind. This is how we come to the judgment that the individual has a disease that has resulted in a disability. And because that individual is a member of the human natural kind, we recognize in that individual an intrinsic value that we call "dignity."

The good at which health care aims is not purely instrumental, but rather, it is a response to our recognition of the intrinsic value of the human. It is in recognition of this worth that we have established the healing professions as our moral response to our fellow humans suffering from disease and disability. The plight of the sick and the disabled rarely serves the purposes, beliefs, desires, interests, or expectations of any of us as individuals. In the end, it is because of the intrinsic value of the sick and disabled that we serve them. Intrinsic human dignity is, thus, the foundation of health care.

11.5 Differences and Natural Kinds

I noted at the beginning of this essay that biological natural kinds are extraordinarily variable, and that this is no less true of the human natural kind. Having explained the concept of natural kinds and its relationship to the notions of disease and dignity, I can now state that this variability is one of the law-like generalizations and typical features and history of the human kind. Part of this variability is genetic, and this genetic variability, over evolutionary time, is vital to the continued flourishing of the kind.

Another law-like generalization and typical feature and history of the human natural kind is our mutual interdependence. This feature is so characteristic and important that MacIntyre (1999) has dubbed human beings "dependent rational animals." We are not merely social in the sense of exercising an option to be gregarious, we are relational beings who need each other. Our flourishing as individuals is, in part, constituted by the common flourishing of the kind.[2]

Given this mutual interdependence, the intrinsic value of the human, and the substantial variability within the kind, it seems prudent to try to classify the various types of differences in individual characteristics in a way that will help us sort through our moral responsibilities to each other. Having grounded human dignity in the bare fact of being a member of the human natural kind, one need not fear that such an exercise will undermine dignity, equality, or rights. Differences can make a difference without threatening one's fundamental moral status once one has accepted the notion that one's fundamental moral status is based upon what one most radically shares in common with all other members of society (i.e., one's humanity) rather than basing one's moral worth on the odd notion that differences make no difference.

I propose three classes of biological differences among members of the human natural kind that are relevant to discussions of health care, rights, opportunity, and social justice. These are (1) variations, (2) differentiations, and (3) diseases, injuries, and disabilities.

11.5.1 Variations

Some differences notable in a biological natural kind are, of themselves, totally irrelevant to the flourishing of the individual, and, if universalized, would not inhibit the flourishing of the kind. Among members of the human natural kind, these include the color of eyes, hair, and skin; whether the individual has an ulnar as well as a radial artery; whether one is right-handed or left-handed. I call these variations. There should be no moral basis for differential access to health care or social treatment of human beings based upon variations.

11.5.2 Differentiations

Some biological differences are actually necessary (in the absence of technology) for the flourishing of a biological natural kind, but do not inhibit the individuals who are characterized by these differences from flourishing as members of the kind. The test criterion is that these differences could not be universalized without inhibiting the common flourishing of the kind. Such differences include, in most animals, male and female sex; in bees, the differentiation into queens, drones, and workers; in mammals, developmental stages such as newborn and pubescent. Among the human natural kind, differential treatment of individuals according to their differentiated states is morally justified only to the extent that it contributes to the flourishing of the kind. Accordingly, one can justify a prohibition on automobile driving by young adolescents, but could not justify denying such individuals, as a class, access to health care. One could justify prohibiting pregnant women from working in factories that produce the drug thalidomide, but could not justify denying women the opportunity to serve as president of a republic.

11.5.3 Diseases, Injuries, and Disabilities

Some differences among members of biological natural kinds, of themselves, inhibit the possibilities for the affected individual to flourish as the kind of thing that each is. If a particular difference fits the definition of disease offered above (or its variation, the definition of injury), then it can be classified in this grouping of biological differences. Even those conditions that are a side-effect of a biological difference that contribute to the flourishing of the kind, but are harmful to at least some individuals (i.e., the diseases that contribute to an "inclusive-fit" advantage) are diseases because they inhibit the flourishing of affected individuals.

11.6 Intrinsic Dignity as the Basis for Human Rights

Intrinsic dignity is, as the UN Declaration on Rights proclaims, the foundation for all considerations of human rights. I have argued elsewhere (Sulmasy, 2007) that the respect that we owe to those who bear dignity implies two kinds of rights: universal human rights and local, stipulative rights.

By universal human rights, I mean those rights that must always and everywhere be respected, binding individuals never to transgress them. Roughly, these will correspond to so-called "negative rights."[3] In Kantian language, these correspond to duties of perfect obligation.

By local, stipulative rights I mean those rights that can be granted by various societies according to their particular means and particular conditions. Roughly, these will be so-called "positive rights"—rights to be given particular goods or services. In Kantian language, these correspond to duties of imperfect obligation.

Perfectionist theories of ethics hold that the goal of morality is to promote human flourishing. The human good consists in the flowering of the dispositional properties that make us human, holistically integrated within each of us and socially integrated in the life we share with others as an inherently social, mutually interdependent natural kind.

Universal human rights proscribe our acting in ways that would deny human beings their intrinsic dignity or prevent them from pursuing the values that they attribute to themselves or others, provided that the pursuit of these attributed values contributes to human flourishing in a fully integral sense. Local, stipulative rights permit access to goods and services that societies (to the extent that physical and social conditions allow) make available in order to foster conditions that are conducive to human flourishing in a fully integral sense. Societies are obligated to grant such local, stipulative rights, as conditions allow, in order to promote human flourishing and are limited in their granting of any such right if it would detract from human flourishing.

Universal rights thus are rights because their violation is inconsistent with human flourishing. Local, stipulative rights thus are established on the basis of their capacity to promote actual human flourishing.

11.7 Rights and the Care of the Disabled

On this theory, certain rights relevant to healthcare are universal human rights and are therefore absolute—e.g., the right not to be killed or tortured by physicians; the right not to be experimented upon without consent; the right not to be cloned. These are the universal human rights directed towards respect for the intrinsic value of the human natural kind. These are negative rights. These are natural rights. They must be respected for the disabled as well as for the able.

On this theory, however, the provision of health care would *not* be considered a universal human right, as I have defined this term. The provision of the goods and services of health care is neither a negative right nor a natural right, for the able or for the disabled. The failure to provide health care does not, in itself, directly prevent persons from pursing the values they attribute to themselves or others. And under certain conditions, some societies simply will not have the resources to provide health care. Therefore it cannot be universally required of all human beings.

However, the provision of health care ought to be considered an extremely important local, stipulative right. A society that has the means to provide health care to its members has an obligation to establish access to health care as a local, stipulative right because health care contributes to the flourishing of both the able and the disabled. In fact, it may contribute relatively *more* to the flourishing of the disabled. Basic health care is an obvious contribution to human flourishing. It is an important and significant expression of human solidarity, mutual interdependence, and respect for intrinsic human dignity even under historical or social conditions in which it is not scientifically efficacious. However, in the 21st century, when medicine has become highly scientifically efficacious, health care can contribute to human flourishing in other highly important ways—by prolonging life; by treating symptoms; by providing opportunities for other forms of human flourishing. Health care does this by curing, relieving the burden of symptoms, and providing ways for those disabled by disease to cope better or even to find alternative ways of exercising their diminished capacities. Because the flourishing of the human natural kind is integrally social—i.e., the flourishing of the individual consists, in part, in the flourishing of the community—those who have the means to provide health care for themselves have an obligation to share their resources with those who do not. This is how one shows respect for the intrinsic dignity of each—the acts of compassion and benevolence that assist others in the pursuit of attributed values that contribute to fully integrated human flourishing.

11.7.1 Health Care and the Disabled

Disability does not transform a human being into another natural kind. One classifies a person as disabled because one has first picked that individual out as a member of the human natural kind, noting the kind-typical features that the individual does not express. The disabled therefore have intrinsic dignity. Respect for intrinsic dignity would dictate, as argued above, that one recognize the radically *equal* intrinsic dignity of a severely mentally retarded adult and that of a philosophy professor at an Ivy League university. This equality is not based upon the peculiar proposition that no differences mark a moral difference, but upon the proposition that the differences that result in disability do not mark a difference in the *kind* of thing the individual is. No matter how disabled a human being may be, there are no gradations in that individual's being a member of the human natural kind and subsequently no differences in intrinsic dignity. This is the value one has by virtue of being the kind of thing that one is—a member of the human natural kind.

A number of moral conclusions can be drawn. First, respect for intrinsic dignity would prohibit, as described above, euthanizing disabled human beings of any age on the basis of their disability. As discussed above, this practice would violate a universal human right.

Second, the differences among humans that result in disability (illnesses or injuries) command moral attention. To the extent that health care can help such persons to flourish more fully as individuals, they have a *prima facie* claim to health care resources that can never be less than the claim of any other individual with whom they share equal dignity. The duty to build up the inflorescent dignity of human beings—a duty based upon respect for intrinsic dignity—carries with it the notion of the radical equality of the intrinsic dignity of all human beings. Just as skin color, income, education, and social worth ought not be the basis for differential access to health care, likewise disability ought not be invoked as a basis for justifying unequal access to health care.

Yet, as a duty of imperfect obligation, the duty to provide health care to the disabled will have limits even in the wealthiest society. As already discussed, the society must have the means to provide health care in the first place. The extent of this obligation is also limited by the acknowledgement that there is a point of diminishing returns. Those societies (and individuals within societies) that have the means to provide for health care for others need not (and actually should not) apportion such quantities of their resources to health care that the human flourishing of each is diminished because the flourishing of the whole has been compromised by excess diversion of resources to the health care of the disabled. Western societies can hardly claim to have reached that point.

Even in the relative absence of scarcity, however, there will be limits in how far one goes in sustaining the life of a disabled person, just as there will be limits to how far one must go in sustaining the life of any person. These limits include the physical, psychological, social, spiritual, and economic resources of the individual in his or her particular circumstances as well as the limits of a society's resources. Any criterion for deciding upon limits must *not* be based upon the disability in itself, since this would constitute a judgment regarding the worth of the person and violate the principle of equal respect for intrinsic dignity. Rather, such judgments must be based on the same criteria one would use for deciding upon the limits of care for any individual, disabled or not. That is, upon the inefficacy of the intervention, absolute scarcity, or the individual's own judgments about burdens and benefits. Limits based upon judgments of social worth, whether made by physicians or third parties, are inconsistent with the meaning of respect for intrinsic dignity.

11.7.2 Disability, Difference, and Health Care

As discussed above, not all differences are equal. Some differences can justify differential treatment without threatening the underlying basis for the fundamental equality of intrinsic human worth. The disabled are different. These differences tend to inhibit their flourishing as individual members of the human natural kind. But as members of a mutually interdependent natural kind, these differences constitute a claim by the disabled upon those who are not disabled. This claim has limits—accommodations must be reasonable; treatments must not be so expensive as to substantially detract from the common good (a common good that includes the disabled and is enhanced when the disabled are cared for).

It is also, *prima facie*, morally wrong to create human beings with the specific intention of making them disabled (whether deaf or genetic dwarfs or mentally retarded). These conditions inhibit human flourishing, and all things considered, it is better not to be disabled. This is emphatically not to say that disabled individuals, with the proper assistance, could not flourish more than many do now. It is certainly not an argument that the disabled should be selectively aborted or euthanized. In fact, quite the opposite: it is a powerful argument that no human being should be aborted or euthanized because this would be justified only by a denial of the intrinsic dignity that grounds all our moral obligations towards each other. Because one does not require the premise that differences in human characteristics make no moral difference in order to ground the claim for the rights of the disabled, the counter-intuitive corollary that intentionally creating disabled human beings is morally permissible can be avoided without sacrificing a vigorous claim to rights for the disabled. Thus, the notion of natural kinds grounds a sensible alternative to the Standard Civil Rights Formula in developing a theory of rights for the disabled.

11.8 Concluding Caveats

The arguments presented in this essay have been very broad. I have presented what for many readers will be a new way of thinking about ethics, and in a single essay (which has already become quite long) I have only been able to present my arguments in outline form. Some of the parts of this argument require further elaboration and justification. Many detailed steps between the parts of the argument have been omitted for the sake of brevity. Some might complain that the argument is overly complex and requires too many steps. The topic is complex, however, and the analysis ought not be over-simplified. I simply suggest that the argument, in outline, is plausible. The basis for equality of treatment of the disabled is not that differences make no difference, but that equal treatment is based upon what the disabled and the able have most radically in common—their humanity. I suggest that this is a sounder argument for grounding a robust set of rights for the disabled than the Standard Civil Rights Formula.

Notes

1. Nordenfelt (1995, pp. 53–57) calls these goals "vital."
2. In fact, the genetic variability that makes for the flourishing of the kind as a whole also results in a substantial number of individual members of the kind who suffer greatly. The generation of genetic variability comes about via mutation. And many mutations result in disease and disability. The fact that this disease and disability is a consequence of the generation of the genetic variability that is necessary for the flourishing of the human natural kind over generations can transform one's understanding of the mutual interdependence of human beings into a template for a kind of biological justice. Nowhere might this be better exemplified than the example of sickle-cell disease and "inclusive-fit" genetics. It is now widely accepted that that the carrier state of sickle cell disease confers substantial biological protection against malaria, while the homozygous state results in a profound, painful disease characterized by substantial disability. As a mutually interdependent natural kind, we have an obligation, in biological justice, to care for those whose individual disability is a consequence of the genetic variability that contributes to the common flourishing of our kind.
3. The negative/positive distinction is sometimes neither clear nor helpful. See, e.g., Beauchamp (1982, pp. 199–201). This is one reason to seek a new classification of rights such as the one I have proposed.

References

Beauchamp, T. L. (1982). *Philosophical ethics*. New York: McGraw-Hill.

Brody, B. (1980). *Identity and essence*. Princeton, NJ: Princeton University Press.

Garcia, L. L. (2008). Natural kinds, persons, and abortion. *National Catholic Bioethics Quarterly, 8*, 265–273.

Kripke, S. (1971). Identity and necessity. In M. K. Munitz (Ed.), *Identity and individuation* (pp. 135–164). New York: New York University Press.

Kripke, S. A. (1972). Naming and necessity. In G. Harman & D. Davidson (Eds.), *Semantics of natural language* (pp. 253–355). Dordrecht, Netherlands: D. Reidel.

MacIntyre, A. (1999).*Dependent rational animals: Why human beings need the virtues*. Chicago: Open Court.

Nordenfelt, L. (1995). *On the nature of health: An action-theoretic approach* (2nd ed.). Dordrecht, Netherlands: Kluwer.

Rawls, J. (1971). *A theory of justice*. Cambridge, Massachusetts: Belknap Press of Harvard University Press.

Shapiro, J. P. (1994). *No pity*. New York: Three Rivers Press.

Sulmasy, D. P. (2005). Diseases and natural kinds. *Theoretical Medicine and Bioethics, 26*, 487–513.

Sulmasy, D. P. (2007). Dignity, rights, health care, and human flourishing. In G. D. Pintos & D. N. Weisstub (Eds.), *Human rights and health care* (pp. 25–36). Dordrecht, Netherlands: Springer.

Sulmasy, D. P. (2008). Dignity and bioethics: history, theory, and selected applications. In E. D. Pellegrino (Ed.), *Human dignity and bioethics* (pp. 469–501). Washington, DC: The President's Council on Bioethics.

United Nations. (1948). *Universal declaration on human rights* [on-line]. Available at http://www.un.org/Overview/rights.html

Wiggins, D. (1980). *Sameness and substance*. Cambridge, Massachusetts: Harvard University Press.

Wiggins, D. (2001). *Sameness and substance renewed*. Cambridge, UK: Cambridge University Press.

Chapter 12
Public Policy and Personal Aspects of Disability

Patricia M. Owens and Eric J. Cassell

Prologue: Specific rule based programs represent society's response to individuals with disabilities. There is no clearly articulated or widely accepted overarching public policy. Existing programs in the United States have evolved from principles of social justice which (perhaps unwittingly) assign medically impaired persons to administrative categories. If they fit these categories, they qualify for compensation in lieu of earnings from work. Emerging principles view disability as a health state as much determined by environmental, social and personal factors as by medical impairment. Each person's state of health has a different trajectory determined by individual circumstances which may or may not culminate in an inability to participate in social roles, including work.

The Americans with Disabilities Act (ADA) placed responsibilities across society for removal of barriers to participation for persons with disability. The ADA thus made participation in society a primary goal of disability policy and programs. To carry out an enabling process, personal factors must be taken into account.

In this chapter we (1) examine selected social justice principles exemplified by existing work- or earnings-based disability programs, (2) emerging views and concepts of disability, and (3) describe personal characteristics and explain how personal aspects can better inform a disability policy aimed at increasing participation in society by persons with disability.

12.1 Introduction

According to United States census data between forty and fifty million individuals in the United States report some kind of disability.[1] Perhaps the most prominent societal response to working aged persons with disability is payment of compensation in lieu of wages. (Over $50 billion in disability compensation of one type or another is paid per year.) To receive disability compensation, persons must fit into

P.M. Owens (✉)
Department of Public Health, Weill Medical College, Cornell University, New York, USA
e-mail: patowens@ericcassell.com

D.C. Ralston, J. Ho (eds.), *Philosophical Reflections on Disability*, Philosophy and Medicine 104, DOI 10.1007/978-90-481-2477-0_12,
© Springer Science+Business Media B.V. 2010

administrative disability categories based for the most part on loss of earnings and severity of medical conditions. Abstract principles of social justice are embodied in the disability categories or programs that have evolved.

Emerging concepts of disability as a health state influenced by personal and environmental factors suggest that principles of social justice as currently practiced do not reflect the more personal nature of disability. There is a knowable distinction between establishing and fitting individuals into administratively constructed disability categories and understanding how a person's state of health becomes disability (or doesn't). Personal considerations are essential to health assessments, with the potential to assist persons with impairments, capitalize on their abilities and participate more fully in society. Under the ADA premise that a more appropriate societal response to disability is increasing participation of individuals in society, including work, the (personal) phenomenology of disability must be understood and applied. Ethical principles—respect for persons, benevolence and justice—are best served by an individualized approach to evaluating and maximizing function. Individual versus societal responsibilities in relationship to participation can be better determined and assigned when there is a clearer understanding of the personal dimension of disability.

12.2 Current Societal Perspectives and Resulting Work Disability Programs

Social justice is a set of principles employed by a society to provide to groups or their members fair treatment and a just share of society's benefits. The principles relative to disability include:

1. Establishing legitimate categories for publically funded compensation in lieu of wages;
2. Liability for risks taken by a person for the benefit of others;
3. Work disability is an insurable risk that can be protected against by either social or private insurance;
4. Addressing concern of citizens about the cost and potential for misuse of publically funded programs;
5. The provision of certain civil rights for citizens including persons with disabilities, e.g. the ADA.

12.2.1 Legitimate Categories

Social justice is made evident in the identification of legitimate categories for exclusion from a society's expected norms. In western societies individuals of a certain age are expected to work and support themselves as a condition of receiving their just share of wealth and resources. Disability is generally considered a legitimate category that justifies not working and qualifying for compensation in lieu of work.

"Each category must be based on a culturally legitimate rationale for nonpartici-pation in the labor system. . . .. The definitions are also tied to underlying cultural notions about work" (Stone, 1984, p. 22).

The Social Security Disability Insurance (SSDI) and Supplemental Security Income for the Disabled (SSI) programs use rule-based categories to provide dis-ability compensation to persons who cannot work because of a medical impairment. These rules include a carefully constructed assessment process, medical documen-tation of a physical or mental medical condition, and an administrative rating of impairment severity and work capacity. For SSDI and SSI, the evidence must show that a person has an impairment expected to last a year or result in death. Because of this impairment, a person must be unable to engage in substantial gainful activ-ity measured by a dollar amount which is presently $940 (see *socialsecurity.gov* website). For SSDI, people who work pay into a fund to protect themselves against work disability and that fund pays if they meet the definition. SSI is a needs-based program where society pays the cost through general revenue when persons meet the disability definition and also meet income and resource limits.

12.2.2 Risk Assumption and its Consequences

When persons assume certain risks, society or some segment of society is expected to compensate persons for loss resulting from these risks. Two primary examples are Military Service/Veterans Programs and Workers Compensation Programs.

Military and Veterans disability-related benefits are available for service mem-bers who put themselves at risk while protecting national security. Society has the liability (pays the costs) to compensate for "loss of earnings" from illness or injury which results from taking this risk. VA benefits are generally not contingent on work status but the VA disability rating is based on average earnings loss attributable to the severity of the impairment. Military service disability benefits are based on length of time in service and whether the impairment prevents an active service member from performing required duties. Rehabilitation, both medical and vocational, is part of the VA and Military benefit system. Disability ratings are also important to determining access to continuing medical care provided by the Veterans Health Care System.

State Workers Compensation (WC) Programs[2] are another example of programs in which society assigns liability for risks as a principle of social justice. In this case, the risk for workers is an injury or illness arising at the work place. Employers have the liability for injuries and illness which arise in the course of employment. WC, in the 1920s, was the first form of social insurance in the United States (originating with the federal government). Now, programs in each of the United States have dif-fering rules, but all of them require employers to pay medical and disability benefits to their employees who sustain impairment from accidents or illness that occur in the course of or arise from employment (Sengupta, Reno, & Burton, 2005).

Both medical costs and compensation are provided in relationship to the severity of the illness or injury. After maximum medical recovery, it is determined whether the illness or injury is permanent or if temporary, total or partial. Most states use a schedule of payments based on severity (Sengupta, Reno, Baker, & Taylor, 2008).

12.2.3 Insurable Risk

Work disability is seen as a risk whose actual occurrence can be predicted by actuarial calculations (Schultz& Gatchel, 2005, pp. 524, 526). This risk is based on projected disability incidence and duration for individuals and groups in similar risk categories. Levels of risk for work disability are determined by individual characteristics such as age, work skills and health as well as the type of work performed (industries and occupations). Such actuarial practices make it possible for private sector insurers to estimate a price for risk protection. Informed individuals exercise their perceived responsibility for purchasing protection (or not) for this risk. Or, similarly, employers use this disability insurance as an employee benefit to help recruit and retain employees. Private sector insurance programs have greater recognition of the importance of personal and environmental factors influencing disability, but are still limited by contract definitions which are categorical.

Social Insurance for events that prevent earning an income has evolved in the United States, SSDI being the prime example. Social Insurance spreads the risk of work disability across the working population. Payroll taxes from all covered workers and their employers are pooled to support payments to those found disabled under the established definition. Everyone pays according to a wage-related formula and a younger person with less risk of becoming disabled pays the same rate as an older person with greater risk.

There is a notion in both public and private insurance approaches that what economists and insurers call "moral hazard" can be managed. The term moral hazard refers to the fact that insurance can negatively change the behavior of the person being insured. In a *New Yorker* Public Policy Article (August 29, 2005), noted commentator Malcolm Gladwell highlighted the relevance of moral hazard:

> Insurance can have the paradoxical effect of producing risky and wasteful behavior. Economists spend a great deal of time thinking about such moral hazards for good reason. Insurance is an attempt to make human life safer and more secure. But, if those efforts can backfire and produce riskier behavior, providing insurance becomes a much more complicated and problematic endeavor.

12.2.4 Misuse of Disability Categories

While western societies overwhelmingly accept disability compensation as a part of responsible social justice principles, there are concerns regarding overuse or misuse of the disability exemption to work (the moral hazard issue). There are also strongly and widely held views that many persons who fit categorical definitions of disability can work given adequate incentives and support. Therefore, they retain a responsibility to work, should be encouraged and supported in work efforts, and should not be discriminated against in the workplace. Economists worry that persons with disabilities who can satisfy their needs through disability compensation may not be motivated to enter the labor market, especially in view of other environmental and employment barriers (Weaver, 1991). Fears of deception, abuse, symptom exaggeration or malingering have generated vigorous tactics to discover and deal with abuses in both public and private programs (Berkowitz and Hill, 1986; Weaver, 1991; Krause et al., 2001).

12.2.5 Civil Rights and Disability

The ADA states that "the Nation's proper goals regarding individuals with disabilities are to assure equality of opportunity, full participation, independent living, and economic self-sufficiency for such individuals."[3] This law, enacted in 1990 and amended in 2008 to restore some limitations imposed by the judicial system, is the most recent sweeping discussion of society's objectives regarding "persons" with disability.

12.3 Changing Views of Disability

The International Classification of Functioning, Disability and Health (ICF) defines disability as a health-related state. Over the past several decades, the conceptual framework of disability has moved from solely biomedical definitions of this state to wider socioeconomic views as demonstrated by the evolution of the ICF (IOM, 1991).

As a new member of the WHO Family of International Classifications, ICF describes how people live with their health condition. ICF is a classification of health and health-related domains that describe body functions and structures, activities and participation. The domains are classified from bodily, individual, and societal perspectives. Since an individual's functioning and disability occurs in a broader context, ICF also includes a list of relevant environmental factors (World Health Organization, 1999; revised in World Health Organization, 2001).

Increasingly, compensation is seen as only one part of a social justice system to increase participation of persons with disabilities. At the same time, it continues to be accepted that for some persons with impairment, actual work activity is not possible. In those cases compensation may be the primary tool to allow for participation and increase quality of life. Personal considerations also help determine when this is the case.

Participation in society—including the performance of apposite social roles, e.g. work—is now frequently cited as the most desirable outcome of social policy (Brandt & Pope, 1997).

Pro-work support policies are discussed by Burkhauser and Stapleton (2003). They maintain that

> [h]istorically, the federal government's approach to providing economic security for people with disabilities has been dominated by a caretaker approach, reflect[ing] the outdated view that disability is solely a medical issue. A main premise of this model is that people with severe medical conditions are unable to work.... The government, at the insistence of advocates and others, has launched a multifaceted effort to change that...
> (Burkhauser and Stapleton, 2003, p. 398)

These authors cite examples of such social policy efforts and instruments as the Americans with Disabilities Act, the 1998 Individuals with Disabilities with Education Act (IDEA), the 1999 Ticket to Work and Work Incentives Improvement Act (TW&WIIA), administration initiatives such as the Clinton Administration's Presidential Task Force on the Employment of Adults with Disabilities, and the Bush administration's New Freedom Initiative (NFI). Burkhauser and Stapleton also maintain that pro-work social justice policy requires "investment in 'the human capital' of people with disabilities" (2003, p. 399). They cite evidence from a survey of private and government employers which indicate that lack of training and lack of related experience are the main barriers to employment and advancement of people with disabilities (Bruyère, Erickson, & VanLooy, 2000).

In *A Disability System for the 21st Century* (September, 2006), the Social Security Advisory Board proposed an alternate to the present Social Security Disability system. They suggest that the first step in the disability claims process would be an "assessment that focuses on what resources he or she, and society, have available to make it possible for that individual to retain or regain capacity for self support" (Social Security Advisory Board, 2006, p. 14). For those who have an ability to work, the process would then provide rehabilitation, support and ultimately job placement. Those found unable to work on this first assessment or those who cannot work even with services would receive compensation. Advocates for persons with disability stress the criticality of having the compensation alternative readily available for those with the most severe impairments.

Another IOM report on disability maintains "[t]hat Americans as a people should take explicit responsibility for defining the future of disability based on a commitment to fully integrating people with disabilities into community life and to developing the knowledge, technologies and public understanding to support that goal" (Institute of Medicine, 2007, p. 1).

Adopting participation in society as a driving objective in today's views of social justice leads to the question of how we go about increasing participation. Employing the precepts of the ICF as a starting point, and viewing disability as a state of health determined by the interaction of individual, social and environment factors, a large part of the answer is understanding the individual as a person. How *do* we understand the person, if it is through this understanding that social and environmental barriers to participation are to be accommodated? It is also important to note that a person's health state is greatly influenced by encounters with the health care system, the world of medicine. In this next section, we discuss how being a person influences the health state of disability.

12.4 Disability and the Person

Categorical definitions of disability undergird the United States disability programs. Definitions usually relate inability to perform tasks or roles (such as work and education) to the severity of medical condition. The failure of medicine among other institutions to attend to the social dimension of the definition of disability— its slow recognition of the fact that something that is social is also inescapably personal—represents another facet of the tardy acceptance of the personal and subjective aspects of sickness and its effects. As a consequence there remains inadequate attention to how the personal and subjective aspects of an impairment of function (or health state) convert the impairment into an inability to perform a social function.

Certain characteristics of disability clarify its personal nature. Disability is most often associated with the existence and severity of a medical condition. But, with exceptions, the severity of the medical condition alone does not predict disability. For example, some persons who have lost limbs, others who live in pain, and some who have difficulty breathing, continue working while others with similar afflictions consider themselves disabled. This is highlighted by the fact that sometimes persons file for disability not because of a worsening of their impairment but because of a loss of transportation, changes in the work-place, or even a change in a supervisor. The meaning and fear of continuing to work as well as the occurrence of new or worsening symptoms affects whether someone believes continued working has become impossible.

Bodies may experience impairments of function but bodies do not create the meaning or sense of the future associated with impairments, *only persons do*. Bodies may have impairments, but only persons have disabilities (Cassell, 1982). Disability in adults occurs when persons conclude that they can no longer fulfill the responsibilities of an expected social role. Disability continues until persons believe they can again perform their expected roles because their social position has adapted to their performance, role expectations have changed, the social context has changed, or their impairments of function have diminished or changed to allow the fulfillment of social expectations. In this understanding of disability the person is brought to the center of the medical equation. Exploration of and response to these personal events can influence the state of health. Disability may occur as a result of the treatment or the behavior of physicians and other caregivers, not just the functional effects of the disease or injury. Thus, patients can move from the existence of impairment to a state of disability without any change in their functional capacities. As pointed out earlier, we call disability a health state rather than merely another step in the continuum of function because so many facets of the whole person and their interactions with others make it qualitatively different. Impairment may or may not result in a person who is experiencing changing views of himself and his roles and responsibilities. However, even with an impairment of fixed severity, a person's views about the effect of that impairment on roles and responsibilities may change over time, resulting in disability—i.e., an inability to work.

12.4.1 Who is the Person with Disability?

It is common in society and especially medicine these days to talk about the ill person, or the person who is the patient, the person with a disability, or about treating the whole person. Ideas about treating the whole patient trace their origins to the 1950s, when medicine was beginning to move away from a sole focus on the disease, and the slogan "treat the patient as a person" became common. The surrounding society was also changing. The civil rights movement in the United States, the re-emergence of the women's movement, the disability rights movement, as well as an increasing stress on individuality moved persons to center stage. Bioethics, which had become a major force in medicine by the 1970s, helped promote these changes, particularly in its stress on patient autonomy. The social changes since World War II have been part of a greater emphasis on the individual and individualism that has occurred not just in the United States, but in the whole world. Patients have come to play a central role in making decisions about their own health and function.

In the following discussion of what it means to be a person, keep in mind the impact of impairment and disability. A person is a thinking, feeling, acting human individual virtually all of whose behaviors—volitional, habitual, instinctual, or automatic—occur based on meanings and within a context of relationships with others and with self. The relationships with others may be actual others or others incorporated into the internalized meaning structure we call the rules of behavior and the meanings of language. The person has a body which can do some things but not others and to whose enormous range of capacities and inabilities persons become habituated. This view of persons has been partly obscured by the cultural importance of and attention to individualism—self-determinism—which has developed over the past number of centuries in Western European and American societies, but it is vitally important in understanding functional impairment and disability.

Persons are always in relationship to other persons, institutions, and society. You will *never* see just-a-person, an isolated-in-a-vacuum-person, because there is no such thing. The atomistic person is as much a myth as the atomistic fact of positivism. The extended web of human relationships and the rules that guide them are called society and culture ("a historically transmitted pattern of meanings embodied in symbols, a system of inherited conceptions expressed in symbolic forms by means of which men communicate, perpetuate, and develop their knowledge about and attitudes toward life") [Geertz, 1973, p. 89]. These relationships to others dominate life at all ages and establish persons' place in society and culture. They determine a person's roles and responsibilities in society. Bruner has called the impact of these factors on mental life "folk psychology" (1990, pp. 33 ff.).

We wrote earlier of how social justice is based on society's expectations of individual rights and *responsibilities*. The ADA clearly sets out responsibilities. However, currently the word "responsibility" is used much more frequently than in the past, and in different situations with sometimes very different meanings. For example, the Governor of New York State, George Pataki, said that "[w]hen Government accepts responsibility for people, then people no longer take responsibility for themselves. Individual responsibility and personal freedom are inevitably linked" (1995). To understand this and its relation to disability, it is helpful to turn to the origins of the word "responsibility."

Examining the etymology, "responsibility" comes from "responsible" which comes from "response" (OED 2nd ed. electronic). Early on "responsibility" implied listening to, then responding, then being accountable for the response and thus *responsible*. Then, the concept of being responsible became what the individual who is responsible *takes on him or herself*. The concept of responsibility would not contain the ideas of *listening to and responding* if there were not some relationship between the one who speaks and the one who listens. The unspoken speaker to whom the individual listens and responds, one can only speculate, is the social group. This is, however, a social group whose relationship to the individual was very different in the 17th century than now—the individual then was part of the body of the group only approximated currently by the position of a child in a family. As the person has become more individualized, the group—now writ large as the employer or government—has become more responsible. Thus, Governor Pataki's expressed fear that shifting responsibility to government will remove it from the individual really expresses the political struggle accompanying the liberation of the individual. But the tug of war is important in thinking about disability. It isn't responsibility that individuality wants to shed but barriers to the exercise of personal responsibility. In the terms of the ADA we (along with the person involved) must identify and remove (if possible) the barriers to the exercise of responsibility. Reducing barriers to the exercise of personal responsibility allows persons to more fully participate in society. With this in mind, we now can look at the changing nature of society's definition of and response to disability in terms of the historical process of the dialectic of individual and society.

No one part of a person is isolated from other parts. To say that persons have bodies also implies that what bodies can and cannot do that is manifest over time and at any moment in time will be the subject of commentary from within the stream of thought that accompanies every waking moment, and equally by the stream of emotion that acts as an evaluatory commentary on both the occurrences of the life from moment to moment as well as on the thoughts.

Let us now add back in the condition that underlies disability. When sickness or accident occur and the capacities of the body are changed, the change may be seen as inconsequential (or temporary) and the person closes around the impairment, adapting and adjusting to it so as to cause as little disturbance to the relationships of persons to others and to themselves as possible. Homeostasis, the tendency of a system to maintain internal stability in response to situations or stimuli tending to disturb its normal function does not just take place at a cellular level. An impairment that reduces functional capacity is potentially a threat to an individual's place in the family ("Is dad still the breadwinner?"), the person's self-image as a man or woman, and the person's economic well-being. Sometimes the degree of physical incapacity that can be accommodated to preserve stability is enormous.

Yet these facts about persons do not provide a complete definition. The boundaries of persons as people are not quite the same as their physical boundaries. That is because the past and the future (which are changing all the time) are part of persons. Because of this the impact of impairment or of disability achieves its meanings from the history of the individual, the history of the individual's family, as well as the larger social group to which the individual belongs. The future is involved because persons tend to project into the future what is happening at the moment and this is as true of impairments and accepting the category of disabled as of other events. It is interesting in all this to see disability and its personal and societal response as a lens through which to view the evolution of the individual and his or her relationship to self and society.

When the primary social justice principle is to support working-age persons in their move toward a goal of participation, it is essential that persons *as persons* are the central consideration. The ethical principles—respect for persons, benevolence and justice—can be best served by an individualized approach to evaluating function and then assigning responsibility for improving and maximizing function through appropriate personal and societal interventions. As disability programs are developed that promote participation, they will depend on evaluation of persons with impairments in such a way that who they are, their background, personal history, work history, education, physical skills, motivation, desires, concerns, and needs all enter into the planning for rehabilitation, financial support, professional assistance and other interventions in order to return them to the work force or other participation in society.

Epilogue: The following are some suggestions for strategic actions that can help better align public disability policy and programs with individuals as persons to increase participation in society.

1. *An organizing authority which accepts the responsibility to convene stakeholders to agree on the policy and set out guiding principles for identifying existing and new programs to carry out this policy in terms of outcomes, administrative feasibility, and costs.*
2. *Identify and convene key stakeholders (especially persons with disabilities) to articulate a clear, concise overarching disability policy. (It can be argued that the ADA has stated this policy in terms of participation and the ICF has established principles for evaluating and intervening in health states to facilitate participation.)*
3. *Protection for persons with disabilities currently receiving services and compensation so that existing benefits are not taken away unless replaced by a personally acceptable alternative.*
4. *Assistance and funding for current disability programs to adapt (when administratively and fiscally responsible) to more personal evaluations in terms of achieving greater participation for individuals.*
5. *Action both by the legislative and executive branches to establish legal frameworks, assign administrative responsibility and fund efforts will be required.*

Notes

1. US Census Bureau, Summary of Health Statistics for U.S. Population; National Health Interview Survey 2006; Summary of Health Statistics for U.S. Children, National Health Interview Survey 2006, U.S. Census Bureau.
2. Before workers' compensation laws, employees seeking compensation for a work-related injury had to file a tort suit and prove that their employer's negligence caused the injury. Employers had three common-law defenses and employees often did not get damages but

sometimes could receive substantial amounts of money if they were successful. Both employ-ers and employees favored reform that would provide predictable compensation regardless of who was at fault but the principle emerged that employers were liable for providing care and compensation for most workplace injuries.

3. Americans with Disabilities Act of 1990 (Pub. L. 101-336) (ADA), as amended, as these titles appear in volume 42 of the United States Code, beginning at section 12101.

References

Berkowitz, M., & Hill, M. A. (Eds.). (1986). *Disability and the labor market*. New York: ILR Press.

Brandt, E. N., & Pope, A. M. (1997). *Enabling America*. Washington, DC: National Academy Press.

Bruner, J. (1990). *Acts of meaning*. Cambridge, MA: Harvard University Press.

Bruyère, S., Erickson, W., & VanLooy, S. (2000, Autumn). HR's role in managing disability in the workplace. *Employment Relations Today, 47*–66. Available at http://digitalcommons. ilr.cornell.edu/edicollect/119

Burkhauser, R. V., & Stapleton, D. C. (2003). A review of the evidence and its implications for policy change. In R. V. Burkhauser & D. C. Stapleton (Eds.), *The decline in employment of people with disabilities* (pp. 369–406). Kalamazoo, MI: W. E. Upjohn Institute.

Cassell, E. J. (1982). The nature of suffering and the goals of medicine. *New England Journal of Medicine, 306*(11), 639–645.

Geertz, C. (1973). *The interpretation of cultures*. New York: Basic Books.

Gladwell, M. (2005). The moral-hazard myth. *The New Yorker* (August 29). Available: http://www.newyorker.com/archive/2005/08/29/050829fa_fact

Institute of Medicine. (1991). *Disability in America: toward a national agenda for prevention* (A. M. Pope & A. R. Tarlov, Eds.). Washington, DC: National Academy Press.

Institute of Medicine. (2007). *The future of disability in America* (M. Field & A. M. Jette, Eds.). Washington, DC: National Academy Press.

Krause, N., Frank, J., Dasinger, L., Sullivan, T. J., & Sinclair, S. S. (2001). Determinants of duration of disability and return to work after work related injury and illness: Challenges for future research. *American Journal of Industrial Medicine, 40*, 464–484.

Oxford English Dictionary (2nd ed.). Electronic version (1989) Available at www.oed.com

Pataki, G. (1995). The Governor: The people seek change: Transcript of Pataki's inaugural speech. *New York Times* (January 2, 1995).

Schultz, I. A., & Gatchel, R. J. (Eds.). (2005). *Handbook of complex occupational disability claims: Early risk identification, intervention, and prevention* (pp. 524, 526). New York: Springer.

Sengupta, I., Reno, V., Baker, C., & Taylor, L. (2008). Report: Workers' compensation in California and in the Nation: Benefit and employer cost trends, 1989–2005. *Workers' compensation brief – April 2008*. Washington, DC: National Academy of Social Insurance. Available online at http://nasi.org/publications2763/publications_list.htm?cat=reports

Sengupta, I., Reno, V., & Burton, J. F., Jr. (2005). *Report: Workers' compensation: Benefits, coverage, and costs*. Washington, DC: National Academy of Social Insurance. Available Online at http://nasi.org/publications2763/publications_list.htm?cat=reports

Social Security Advisory Board. (2006). *A Disability System for the 21st Century*. September, 2006. Available: http://www.ssab.gov/documents/disability-system-21st.pdf

Stone, D. A. (1984). *The disabled state*. Philadelphia, PA: Temple University Press.

Weaver, C. (1991). *Disability and work*. Washington, DC: University Press of America.

World Health Organization. (1999). *ICIDH-2 International classification of functioning and disability – Beta-2 draft*. Geneva, Switzerland: World Health Organization.

World Health Organization. (2001). *Introduction to the ICF* [On-line]. Available at http://www. who.int/classifications/icd/icdonlineversions/en/index.html

Chapter 13
Disability and Social Justice

Christopher Tollefsen

13.1 Introduction

The core of liberalism, as a political philosophy, involves the recognition that human persons are free and equal, and that the state and its activities should respect these two correlative features of persons. The way in which these features are to be respected varies, of course, sometimes radically: think of the difference between what it means for Locke, and what it means for Rousseau, for a state to respect the freedom and equality of persons (see Locke, 1986; Rousseau, 1997). But this core recognition is at the basis of both Locke and Rousseau's thought; it is similarly at the basis of Kant's political thought (Kant, 1970); and it continues to play an essential—even the essential—role in the liberal political thought of the twentieth century, especially in the work of that century's preeminent political theorist, John Rawls (Rawls, 1971, 1996).

The gains of liberalism are in many respects obvious, important, and indisputable. Yet in the past ten or so years, a form of criticism has been articulated that charges liberalism, both in its origins, and in its developed forms, with an inability to deal with a significant aspect of social life, namely, the facts of disability and dependence. So, for example, have thinkers as diverse as Alasdair MacIntyre, Eva Kittay, Hans Reinders, and, most recently, Martha Nussbaum, all argued (see MacIntyre, 1999; Kittay, 1999; Reinders, 2000; Nussbaum, 2006).

The source of their disquiet is in the interpretation that has been placed on the phrase "free and equal," particularly as that interpretation has been seen through the lens of a third trait frequently lauded by liberalism, independence. Especially insofar as devices such as the social contract, or the original position, have been used to model the relations of free and equal persons, it has seemed that what liberalism *means* by the phrase is: individuals with the active abilities and dispositions to assert moral claims, and to engage reciprocally with those others who also assert moral claims to their mutual advantage. So, for Locke, free and equal individuals

C. Tollefsen (✉)
Department of Philosophy, University of South Carolina, Columbia, South Carolina, USA
e-mail: Christopher.Tollefsen@gmail.com

D.C. Ralston, J. Ho (eds.), *Philosophical Reflections on Disability*, Philosophy and
Medicine 104, DOI 10.1007/978-90-481-2477-0_13,
© Springer Science+Business Media B.V. 2010

are precisely those capable of entering into the social contract; for Rousseau, those individuals capable of entering into and being guided by the general will; and for Rawls, those individuals capable of being "fully cooperating members of society over the course of a complete life" (Rawls, 1980, p. 546).

On any such interpretation, however, the radically disabled—those whose lives as a whole, or for some significant part, are characterized by continued and extreme *dependency*—will be outside the social contract. Their needs will be opaque, and concern for those needs at best an afterthought—a part of the theory to be worked out later, if at all. As Nussbaum points out, the contract theory does not adequately distinguish between those who enter into the contract, or frame its principles ("by whom") and those who benefit from the terms ("for whom"). Rather, it is assumed for the most part that those who contract do so for their own sakes (Nussbaum, 2006, pp. 14–18).

When the oversight is corrected, then it can be seen that, while at least some of the disabled who would not ordinarily be recognized as potential partners to the contract could be empowered so as to become partners, still there are those more profoundly disabled who will never be parties to the contract, and for whom the question of benefits of the contract are therefore likely to seem extra to the concerns of justice proper.

Moreover, as Kittay especially has argued, insofar as the model person for liberalism is one who pursues his or her ends as a "self-originator" of moral claims, and enters into relations with others as a matter of the reciprocity due those others as free, equal, and ultimately independent, the inevitable concern that some have *for* the radically disabled will be left out of consideration in deliberations about justice. For while care for the disabled can be and often is a source of fulfillment for the caregiver, the claims which the caregiver must attend to in the first instance are precisely the claims *of the dependent* (Kittay, 1999, pp. 94–96). Or, as Nussbaum suggests, when the independence of the contracting agent is the focus, important issues of justice such as "the allocation of care, the labor involved in caring, and the social acts of promoting the fuller inclusion of disabled citizens" are deferred or even ignored (Nussbaum, 2006, p. 33).

There are various ways of responding to these difficulties. In this chapter, I am primarily concerned with *political responses*; yet I note first a seemingly metaphysical response with political implications. This response involves a rejection of one ontological characterization of disability as, essentially, a medical problem, in favor of a characterization of disability as exclusively, or nearly so, an issue of social discrimination. If disability is an entirely socially constructed reality, and is constructed by way of the indifference or active hostility of the majority to the minority who are different, then liberal justice can be reconceived on a model of justice for other minorities, such as ethnic minorities and women. By seeing the problem of disability as akin to the problem of race, the social model of disability seems to make that problem more tractable to liberalism (Silvers, 1998).

But other responses move more to the assumed framework of liberal political theory. One such approach to these difficulties is to reframe the nature of the liberal state, typically by moving beyond the contractarian features that seem primarily

responsible for the narrow interpretation of "free and equal." Nussbaum's sup-
plementation of Rawls' theory of justice with the capabilities approach, with a
foundation in the idea of human dignity, and with an Aristotelian conception of
the person, is an example (Nussbaum, 2006; 2000). A second approach is to concede to liberalism at the political level many of the negative consequences for the
disabled, and attempt to build a social ethic capable of addressing the needs of the
dependent and disabled more satisfactorily than liberalism, but at something other
than a political level. In different ways, it seems to me that MacIntyre and Reinders
adopt this approach (MacIntyre, 1999; Reinders, 2000). Finally, more radical criticism is possible, criticism that holds that liberalism is in some fatal way intrinsically
flawed. Kittay, with her principle of *doulia*, or care, perhaps moves in this direction,
although she too stays ultimately within a liberal framework (Kittay, 1999).[1]

All these approaches, including the radical social model of disability, point to
important concerns for a theory of social justice and disability. In particular, there
are four: two concerned with the disabled, and two with those whom I will call,
following Kittay, the dependency worker.

First, as the social model is correct to point out (and Reinders also emphasizes),
the *attitudes* taken towards the disabled by society constitute in their own right both
a constitutive element of justice or its lack, and a source of further justice (or its
lack). A society richly indifferent to, or hostile toward, the disabled is by that fact
unjust and is thereby likely to construe its own terms of existence in ways hostile to
the well-being of the disabled.

But, second, the disabled themselves, even apart from the environment in which
they live, require in many cases active assistance to make possible the flourishing
they may enjoy, as both Kittay and Nussbaum stress. Concern for justice therefore
seems to require going beyond merely a critical approach to the hurdles society
can place against the disabled, to a question of what further steps must be taken.
(Strong defenders of the social model of disability dissent from this claim; see
Silvers, 1998.)

On the part of the dependency worker too there are two concerns that parallel
those of the disabled. On the one hand, there is the attitudinal issue: a society which
does not have adequate respect for the labor of the dependency worker seems both
constitutively unjust, and more likely to throw up hurdles to that labor (as Kittay
argues the Clinton administration's welfare reform bill did). On the other, dependency workers are frequently not given the special help they need in order to be as
capable as they might be of full human flourishing, given the demands they must
satisfy as such workers.

In this chapter, I take a slightly different approach than the modern liberal one.
Rather than beginning, as liberalism does, with the question of how to model a
politics around a (particular) conception of freedom and equality, I begin instead
by asking what the purposes of the state are, and arguing that the state exists to
serve a set of human needs, needs understood by way of a particular account of
human flourishing. This teleological, or natural law, conception of the nature of the
state, I will argue, suffices to justify—indeed, require—considerable concern for
the dependent and disabled and for the dependency worker of the four types just

identified. Each type of needed concern can be worked into the fabric of the natural law approach to politics; and, as I shall show, the foundation for these kinds of concern is *the same* as the foundational concern that generates the need for a state. Thus it is the same foundational concern that generates considerations regarding all other persons within the state.

I then address the question of freedom and equality, the question from which liberal political theory takes its starting point. The natural law approach, I argue, can offer a twofold interpretation of freedom and equality, one metaphysical, one political. Distinguishing the two interpretations in turn prevents a potentially distorted focus on citizenship, a distortion that is manifested in one way, under pressure of the social contract, by an exclusion of the disabled from the scope of the state's proper concern. Thus, on the natural law view, the disabled are not an afterthought, nor is concern for the disabled extra to justice. The distortion also manifests itself in a need to see all disabled persons as capable of citizenship in order to be capable of some measure of human well-being. The natural law view need not hold, however, that citizenship is a fundamental need of the disabled.

13.2 The Problem of Self-Sufficiency

Human deliberation, choice, and action all have as their point human flourishing and well-being.[2] And practical reflection on the well-being possible through action to deliberating agents terminates in a finite number of basic goods, aspects of human well-being recognizable to human practical intelligence as desirable in themselves, and thus as underwriting the desirability of possible states of affairs which could be brought about through action. Such goods include life and health, knowledge, aesthetic experience, work and play, friendship, marriage, harmony with God, and the variety of forms of harmony possible within one's own complex person: harmony of choice and action; judgment and choice; judgment, choice, action, and emotion.

In pursuing these goods, human beings need guidance, since there is a multiplicity of goods, incommensurable each with each. Maximization is not an option, but an attitude of openness to the goods is, and is prompted by practical reason. Most generally, such a normatively required attitude is expressed in the most general principle of morality: act always with a will compatible with integral human fulfillment. This general principle in turn can be specified further, as practical reason identifies ways of willing not so compatible, e.g., directly willing (intending) damage or destruction to one of the goods. Further specification may bring this norm to bear on an action description that makes it clear that a proposed action does involve such a will: the intended death of a patient in mercy killing, say. The norm can thus be applied in the particular case (here, there is an intervening specification: no intentional killing of persons).

This summary of the first principles of a natural law ethics is extremely brief, but is intended primarily to establish the boundaries at which political questions arise. My proposal is that they arise at a particular juncture where the lack of self-sufficiency of human persons for flourishing is apparent.

Individuals just on their own are insufficient for their own flourishing: they require friends, marriage requires a spouse, and even substantive goods such as knowledge and aesthetic experience will suffer in the absence of cooperation and the generation through time of social forms and practices aimed at pursuit of these goods. So a flourishing human life is necessarily communal in various respects. It requires families, networks of friends, and cooperative social structures for the pursuit of goods. Pursuit of the good of religion, too, is typically communal, and, in developed forms with traditions of revelation, awareness of the full range of options of understanding and worship requires access to the tradition in the form, e.g., of an established church. In all these ways the inadequacy of individuals for human flourishing is answered by pursuit of social goods, and the establishment of social institutions that serve both individual and social goods.

Consider, though, a pre-political society, or, more realistically, an overlapping set of such societies: multiple families, social institutions, churches, forms of work, businesses, and so on. Considering them as pre-political is, of course, an abstraction, and no state of nature argument is intended. The point, rather, is to show that although, as regards *what* goods are being pursued, persons in such an overlapping set of societies *are* self-sufficient, they are nevertheless *not* self-sufficient in a number of more instrumental ways.

In the move from consideration of individual pursuit of goods to social pursuit of goods, there were two kinds of inadequacy, two ways in which individuals were not self-sufficient. First, they were not self-sufficient as regards the range of goods they could pursue—social goods cannot be pursued except socially. Second, they were not self-sufficient as regards their *effectiveness* at pursuit of various goods that could only be pursued well socially. The first form of non-self-sufficiency is no longer present in consideration of the overlapping societies—all the goods are now being pursued. And the second too seems not present, or at least not in the same way: we are imagining a state of affairs in which groups really do pursue, e.g., knowledge together, both at a time, and through time.

Yet this set of overlapping social realities is manifestly inadequate for human flourishing in the following ways. First, they are subject to external attack from those outsiders who wish to take their resources or otherwise wrong them. Second, individuals are subject to wrongs internally by free-riders, and violators of distributive and commutative justice. Third, the various groups together suffer from a huge number of coordination problems, a lack of a common and accepted way of doing things together and separately as regards, e.g., transactions of goods, transportation and travel, resolution of disputes, and so on. And finally, and, in the context of this chapter importantly, there is a further kind of lack of self-sufficiency.

Everyone begins life, most end life, and many spend additional periods of their life, in conditions of extreme dependency. Within the overlapping web of pre-political social realities envisaged here, many of those in such conditions of dependency are cared for by those with natural obligations to provide such care. In particular, families owe care to children, to the aged, and to the disabled; neighbors have some obligations as well. Moreover, some of the further social groups, such as Churches, also take on such obligations, as part of their self-understood mission.

But the overlapping set of social realities is inadequate to all the problems of dependency and disability for at least the following reasons. First, not all individuals whose lives are so characterized *do* already, or continue to, live within small social structures in which the obligation to care is recognized and met. Some are orphans, others are parents whose children have died; it is not clear *whose* obligation it is to provide care, and perhaps it is no particular person's obligation. Thus there is a threat that the dependents' needs will not be met.

Second, some who do have specified obligations to the dependent renege on those obligations, or otherwise abuse their charges. Again, the needs of the dependent are threatened.

Finally, some (and probably many) who make good faith efforts to meet the needs of those to whom they have obligations are incapable of fully meeting those needs. Here again, the needs of the dependent are threatened, as additionally are the needs of their caregivers, whose needs include both the care they provide, and other needs that might go unmet given the expense of resources on behalf of their charges. Both the well-being of the dependent, and of those who care for them are thus jeopardized.

13.3 Political Authority

I will return to these three points shortly. At the moment, however, I wish to draw attention to the way in which these four sets of inadequacies—these four ways in which individual and social groups are not self-sufficient—generate the need for a specifically political social reality. All four difficulties may be described as problems of justice and peace – the peace of the overlapping set of communities and social groups is threatened by external and internal force, and by the lack of coordination amongst all members (see Finnis, 1998, p. 227). Justice for these groups and their individual members is threatened also by violence and fraud within, by the inability to punish fairly, and by states of affairs in which the needs of some of the dependent or their caretakers are not met, while the needs of others are (or, in the most extreme cases, when the needs of no dependents are met). But the overlapping sets of social realities themselves are incapable of addressing these problems of justice and peace because they do not have any overarching authority capable of providing definitive and binding solutions to these difficulties. There is a need, in other words, for political authority.[3]

Now, this authority may come to exist simply because some person or group has taken upon themselves the responsibilities of authority and are in fact followed; there is no myth of consent undergirding the picture. But there are clearly forms of authority more and less adequate to the initial needs, and to the condition of the persons with those needs. The authority must, for example, use the coercive force of the sword, and make judgments regarding the good and bad, right and wrong, guilty and innocent within the overlapping set of societies. Such power may be used arbitrarily or selfishly. So an ideal of authority can develop in which the agents of authority are themselves governed by the impersonal authority of law, and, even further, in which those agents are at the same time, both authorities *and* subjects, who take their turn ruling and being ruled.

We are here quite close to the ideals of liberal democratic politics. But those ideals, insofar as they are liberal and democratic, enter into the picture later than the need for political authority just as such. The needs that govern the creation of political authority are the needs of all human beings within the set of overlapping communities, including those proximate, but for whatever contingent reason not currently cared for by some particular community; call these needs *human needs*, or *needs of flourishing*. The ideals of liberal democracy, and of democratic citizenship, are not foundational needs for political authority, but a constraint on how that authority most reasonably should be constituted. There is a need for such constraints, and for a democratic mode of politics; but it is a *need of citizenship*.

The distinction might seem somewhat artificial; but I believe it is of help in solving some of the most serious deficiencies of liberalism as regards the disabled. To show this, I first show that the picture I have drawn accommodates all four concerns outlined earlier as regards both the disabled and the dependency worker. I then show that the natural law approach resolves, for the disabled, some of the key difficulties raised by the social contract approach. Finally, I turn to the questions of the disabled and citizenship, and the disabled and abortion.

13.4 The Needs of the Disabled and the Dependency Worker

Two of those concerns noted earlier were, to reiterate, that both the disabled and the dependency worker had needs for flourishing which they, and those around them, could not meet—respects in which they and their social world were inadequate, not self-sufficient. The very lives of the profoundly disabled are jeopardized by inadequate care; and their ability, and the ability of the seriously, but not profoundly disabled, both mentally and physically, to pursue those goods they are capable of pursuing—knowledge, friendship, play—are jeopardized as well by inadequate care and education. Similarly, the flourishing of the dependency worker was threatened by the expenditure of time and resources that good care for their charges required.

We must note that life, knowledge, friendship, play, and aesthetic experience are real goods for the disabled; Eva Kittay's description of Sesha, her daughter, who cannot even speak, makes clear that Sesha nevertheless suffers when she is ill, thrives in the presence of those who love her, and is benefited by being able to listen to music and watch skilled performances on television. Martha Nussbaum's description of her nephew, Arthur, who has both Asperger's syndrome and Tourette's syndrome, makes clear what an accomplishment it is, and how beneficial for him, when he begins, after two years of education, to recognize the importance of inquiring into the well-being of his friends. And even those so dependent that they are no longer capable of responding to the love and care of family members, such as those in a persistent vegetative state, are better off for being loved and tended, than if they were to be abandoned, mocked, or terminated.[4] For the profoundly disabled, the possibilities of flourishing are different— and in many ways, it must be admitted, limited—relative to the possibility of flourishing for those who are fully capable; yet they are genuine possibilities nonetheless. The dependent and disabled in question are, after all, human beings, and the goods are human goods.

But the ways in which the dependent and disabled are unable to pursue these goods adequately, or as adequately as they might be able to, are continuous with the ways in which all human beings, at various times, are unable to pursue these goods; and they are the very same goods. So the needs of the dependent and disabled stand at the foundation of political society just as much as any other set of needs for assistance in the pursuit of human flourishing.

This is true too of the needs of the dependency worker, with, I think, one important additional feature. For my purposes here, I will assume that the paradigm case of a dependency worker is that of a family member, with obligations to care for a relative—whether child, spouse, or parent, in some condition of extreme dependence; whether temporary or permanent. The resources of such agents are often greatly taxed by such work: the time, money, labor, and emotional investment necessary are considerable, and can leave such persons without the energy or material resources to do much else besides care and rest. (Kittay, for example, notes how beneficial but unusual is her situation, in which she has been able to share burdens both with her academic husband, and with an almost family-like caregiver in her employ.) These demands are threats to the caregiver's well-being, and her ability to ward off those threats is an axis along which she should be considered less than fully self-sufficient. She stands to be benefited as regards the flourishing she is capable of if her family, and social networks, come to her aid. But these, in turn, may require assistance at the political level; the needs of dependency workers are among the needs foundational to the justification for political authority.

Moreover—and here there is a crucial difference, in many cases, between the needs of the dependent and the needs of the caregiver—there is, in the obligations which unexpected dependency create, a threat to the moral well-being of the caregiver. Consider the number of parents who have aborted disabled children, or refused them medical attention at birth, or "put away" disabled children in homes, or neglected elderly parents, or abandoned disabled spouses. Such behavior is to be expected: the agents must shoulder a moral burden for which they are often unprepared, and which they see does not fall on the shoulders of all those around them.[5]

It is a commonplace that the burdens necessary to achieve the common good of a political society should be shared as fairly as possible. To this thought I would add that certain moral burdens should also be shared to the extent possible. Suppose one believes that abortion is a great moral wrong that should be prohibited at law. One can nevertheless recognize the temptation to abortion suffered by those unprepared to bear the moral burden of unexpected care for a child; that burden can be shared, to an extent, by providing the social structures necessary to help unprepared parents care for their children, and the financial and other resources those parents need to be secure in making the right choice. Similarly, the moral burden of all those in positions likely to suffer a strong temptation to abandon, mistreat, or neglect the dependent and disabled can be shared by a societal willingness to provide similar help and resources. In this respect, political society does not *make* the right choices for such potential caregivers, but it can create to some extent a world in which those choices are shielded from adverse consequences; or, more accurately, it can assist others in meeting their responsibility to create such a world.

This leads to the second pair of concerns discussed earlier in this essay: those concerned with the moral, and other social, attitudes towards both the disabled and their caretakers. On the part of the disabled, hostility, discrimination, and indifference result in an environment which is more antagonistic to their flourishing than it need be, or than is reasonable. On the part of the dependency worker, indifference and lack of respect result in a sense that their work is unappreciated, and that they would be better off doing something more socially valued and financially remunerative. How can we see these concerns as aspects of the lack of self-sufficiency on the part of the disabled and their caregivers, such that they are proper concerns of the state?

I hope to suggest the beginnings of an answer by way of analogy. Robert P. George, in *Making Men Moral*, introduced the notion of a moral ecology as part of the common good that a political society exists to serve (George, 1993). The notion fits in with the idea of self-sufficiency and its lack that I have discussed: the ability of an individual to make upright choices is greatly enhanced to the extent that that individual lives in a morally upright environment, and diminished if the situation is otherwise. So, for example, the choice to use or not use pornography is, for competent agents, always a matter of their own will; but they are helped or hindered in making the right choices insofar as pornography is rampant or limited, tolerated or scorned, publicly displayed or only privately consumed. And no individual is self-sufficient to create for himself the sort of environment conducive to virtue.

This type of legitimate concern of the state with the moral ecology of its citizens seems to me to exist at the border of issues of internal commutative injustice—the wrongs perpetrated by one citizen against another—for those who publicize pornography, for example, even if only to sell to other consenting adults, nevertheless show an unjust indifference to the goods of children and families. This concern seems also to exist at the border of issues of internal distributive justice—the need to provide the resources necessary for flourishing to individuals in those respects in which they, or their immediate caregivers, cannot provide them—for parents, like their children, cannot create an entirely reasonable moral environment for their children, except within the home.

The idea of the importance of a moral ecology as a part of the common good makes clear that, insofar as it acts with that good in mind, a state is not only permitted, but sometimes obliged to regulate otherwise private vices for public ends. A morally impermissible but otherwise private taste in pornography becomes a public matter insofar as it helps to create an environment in which the young are corrupted, and parents thwarted in their attempts at child-rearing, even if there was no directly corruptive intent on the part of the purveyors and consumers of pornography.

We can similarly see the attitudes of hostility and even indifference on the part of society towards the disabled and their caregivers as politically problematic because of the social world they create. For the disabled, this world is hostile to flourishing insofar as it creates structural barriers to movement, education, and occupation; these barriers are sometimes a function of indifference, sometimes a function of hostility to the disabled. But the situation can be much worse than mere lack of opportunity, as when special inducements exist for parents to get rid of their disabled children; as when genetic testing is used to help parents abort Down syndrome children; as when laws permit abortion of the disabled up to a later date than laws permitting "elective" abortions; as when medical practice favors starvation and non-treatment of disabled infants; and as when hostility is displayed when the need for publicly funded-services for the disabled is voiced. The currency in society of the idea of a life not worth living—an idea for which we are largely indebted to philosophers—creates for the disabled a world in which their sense of worth and ability to flourish is jeopardized in ways beyond their control to rectify. Just as the state may clear up the public display and sale of pornography for the sake of a moral ecology, so, I think, may it take steps to clear up the array of hostile attitudes towards the disabled, not merely by changing structural features of the environment that are a result of those attitudes (like non-ramped buildings) but also by eliminating some avenues of expression of those attitudes, such as Prenatal Genetic Diagnosis (PGD) for the purposes of abortion, more permissive conditions for the abortion of the disabled, wrongful birth suits, and the de facto immunity of medical practitioners from prosecution for failing to treat "defectives."

On the part of the dependency worker, too, the idea of a moral ecology can play an important role. If we take, again, as our paradigm of the dependency worker a family member who already has obligations to his or her charge, it is clearly unjust for society to permit an environment in which such workers are considered socially unproductive, drains on the economy or, in the case especially of parents who have not aborted disabled children, responsible for their own situation. It is, no doubt, not fully possible for the state to address such attitudes directly, but by changing many of the laws that create an adverse ecology for the disabled, the state can signal its support for dependency workers. Similarly, by providing the resources which distributive justice calls for in regard to dependency workers the state signals to all its concern and takes a step in the direction of creating a more adequate moral environment in which dependency workers can not only fulfill their responsibilities, but also can maintain an adequate degree of the "social bases of self-respect."

13.5 The Disabled and Citizenship

I have distinguished in this chapter between needs of flourishing and needs of citizenship. Societies, and even the pre-political society of overlapping societies that has a need for political authority, all manifest, in each and every individual human being, the needs of flourishing. But the needs of citizenship are needs that arise in the context of the exercise of political authority; they are constraints necessary for government that is congruent with the status of persons.

That status is, as classical liberalism recognizes, a status of free and equal persons. But I would suggest that at this stage of conceptual analysis of the nature of politics, freedom and equality are to be given a political interpretation, whereas in the discussion of the needs of flourishing, freedom and equality should be given a metaphysical interpretation. In other words, the status of beings who may be benefited by participation in the basic goods is the status of persons: beings who are free and rational by nature, and hence equally entitled to full moral consideration. But the beings who meet this description are *human beings*, and, in fact, *every* human being meets this description. Even the profoundly disabled—even those, for example, in a persistent vegetative state—are members of the human species, and to be a member of this species is to have a radical capacity for free action and rational thought, even if, by disease, genetic impairment, or environmental causes, some particular human being or other is rendered unable to actualize that radical capacity.[6] From this perspective, there is no inequality between any two human beings, for all are equally members of this species, a species whose members are, by the fact of their membership, persons.

Yet these claims cannot be unequivocally sustained at the political level. The sense in which adult human beings capable of governing as well as being governed are free and equal is not the sense in which those same human beings are the equals of children, for example, who are not yet ready to be *politically* governed, or to govern politically. Children do not yet manifest the needs of citizens, only the needs of flourishing.

The liberal democratic state in which citizens share in ruling and being ruled is thus a constraint on the *political* solution to the needs of flourishing; it is an answer to the needs of citizenship. But those with these needs are a subset of those with the needs of flourishing, for whom the political solution exists. The empowerments, rights, and responsibilities of citizenship are desirable for many, but they are not themselves among the primary needs for which political authority is needed in the first place. In other words, the state does not exist because people have a need to be citizens; rather, given the need for a state, and political authority, citizenship is a supervening need that arises from consideration about how political authority may most appropriately be exercised.

In her recent work, Martha Nussbaum goes some distance towards a position such as this, I believe, in distinguishing between those by whom the social contract is made and those for whom it is made, and in moving away from the Kantian political conception of the person as a rational and independent chooser. Yet Nussbaum's concerns for the disabled still seem fundamentally framed as concerns of citizenship, and indeed, she seems to equate personhood with citizenship at times. Thus, when she is criticizing a guardianship model, she objects that this model

> ...makes the dependents not full parts of the "we" and the "our," not fully equal subjects of political justice. They are taken into account because some member of the "we" happens to care about their interests, not because they are *citizens with rights, equal ends in themselves* (Nussbaum, 2006, p. 238, emphasis added).

Similarly, Nussbaum describes the capabilities, considered as entitlements, as the "fundamental entitlements of citizens," (Nussbaum, 2006, p. 166); and, later, writes that "citizens enjoy full equality only when they are capable of exercising the whole range of capabilities" (Nussbaum, 2006, p. 218).

What is the consequence of this conflation of citizenship with personhood, and the needs of citizenship with the needs of flourishing? Primarily, it seems to me, that those human beings who cannot be anything but passive as regards their relations with others are no longer considered either citizens *or* persons, and are thus excluded from the concerns of the state and the moral community entirely. So, in describing the relationship between capabilities and human identity, Nussbaum writes,

> If enough of them are impossible (as in the case of a person in a persistent vegetative state), we may judge that the life is not a human life at all, any more...In other words, we say of some conditions of a being, let us say a permanent vegetative state of a (former) human being, that this just is not a human life at all, in any meaningful way (Nussbaum, 2006, p. 181).

And, a bit later, "Some types of mental deprivation are so acute that it seems sensible to say that the life there is simply not a human life at all, but a different form of life. Only sentiment leads us to call the person in a persistent vegetative condition, or an anencephalic child, human" (Nussbaum, 2006, p. 187).

These joint emphases on active capabilities as linked to both citizenship and personhood create difficulties for Nussbaum's treatment of disabled persons for whom she clearly has a great deal of sympathy, such as Kittay's daughter Sesha, who is incapable of speech, independent living, and much else. The following passage indicates both the overemphasis on citizenship, and the peculiar way it shapes Nussbaum's questions about what is owed to Sesha:

> ...even if Sesha cannot become a potential voter, we should ask what other ways there might be to give her political membership and the possibility of some political activity (although we could allow her a vote through a guardian, as a sign of her full political equality). It is clear that citizens with Down syndrome have participated successfully in their political environment. We should ask how we might arrange that Sesha, too, could have some of these functionings available to her. Again, citizens with many impairments are capable of employment. If Sesha cannot hold a job, well, what other ways might there be to give her some measure of control over her material environment?.... Maintaining a single list of capabilities raises all these questions, and they are vital ones, if people with mental impairments and disabilities are to be fully equal as citizens (Nussbaum, 2006, pp. 194–195).

But why must we route these questions through a concern for Sesha's *citizenship*? Sesha is a human being among other human beings, part of a family, and of at least one friendship, and a part of the prepolitical society of overlapping societies, for the well-being of whose members the state exists. That she is not a viable beneficiary of the needs of citizenship in no way jeopardizes her claim to the benefits for which the state exists. In point of fact, Sesha's needs, and the needs of the anencephalic child or the adult in the persistent vegetative state, are, while more extreme, fundamentally like the needs of every other person for whom the state exists in that they are needs of a human being—respects in which some, many, or all human beings are sometimes, often, or always non-self-sufficient as regards their well-being.[7]

These differences point to the underlying disagreement between Nussbaum's conception of the person and mine. For her "Aristotelian" conception of the person is, from the beginning, political: "The political conception of the person that [the capabilities approach] uses includes the idea of the human being as 'by nature' political, that is, as finding deep fulfillment in political relations... The Aristotelian account insists that the good of a human being is both social and political" (Nussbaum, 2006, p. 86).

By contrast, my view has it that while human beings are from the first social— sociality is intrinsic to their well-being—they are political in a secondary way. This way can be described as "natural," certainly, for the need for the political is inevitable to creatures who are social, and the political life, as serving important human needs, can be deeply fulfilling; but it is dependent on that prior sociality, and not on all fours with it. Similarly, it is dependent upon a prior set of basic goods, constitutive of human flourishing, for pursuit of which the political is instrumental.

This means, to reiterate, that on the natural law view described here, the moderately disabled, the temporarily dependent, the "normal" human person, the profoundly retarded, the brain damaged, and even those in a persistent vegetative state, are all alike as regards the fundamental reason that justifies political authority: all are inadequate *in some respect or other* for their own flourishing. All lack self-sufficiency in regards to the conditions necessary for them to achieve the level of well-being they are capable of, *even* when they are situated in particular social networks, and the larger web of overlapping social networks. No special attempt need be made to see any of them as citizens, or potential citizens, or even like citizens, in order to see that they fall within the fundamental scope of the political authority's concern, the basic commitment "to foster the dignity and well-being of all persons within [the state's] borders."[8]

13.6 Some Concluding Thoughts

Much more would need to be said to address all the issues and difficulties my account thus far has raised. But two related points here should be made about the implications of the approach taken, points that I cannot adequately address here.

On this approach, as mentioned (and here again, see MacIntyre, 1999), the disabled and dependent are not, as MacIntyre puts it, a "special" interest, for their interests are the very ones that, shared by all members of the society of overlapping societies, generate the need for political authority. This commonality is one that naturally, however, should direct our attention also to *difference*: the needs, the lacks, the inadequacies, the various ways in which human persons in society, and, indeed, many societies themselves, are non-self-sufficient—all of these are unique, though they display, of course, many common features as well.

This is seen clearly enough in the range of "normal" cases: the life best suited to one agent might not be best suited to another, and even an individual agent will have to reflect upon and discern which of various good options for the shaping of his or her life might be most suitable. But this is true also for the disabled: the particular way of life within which the gifts and capacities of one Down syndrome child will be best pursued and instantiated are not the same for all Down syndrome children. And even the most suitable way of life for *this* patient in a persistent vegetative state—a way of being with his or her family, of being cared for by his or her neighbors and parishioners—might very well differ from the most suitable way of life for *that* patient.

This need for locality in considerations of individual flourishing should indicate both (1) the importance, where possible, of autonomy and freedom for the disabled, but also (2) the fact that, insofar as either the disabled are not, or are not yet, autonomous, and insofar as they have needs with respect to which they are not self-sufficient, their primary caregivers will be family members, friends, and local societies (such as churches)—all of whom are better situated epistemically, emotionally, and volitionally than is the political authority. That is, family members such as parents and spouses, friends, and parishes all have a better grasp of the particular needs and capacities of individuals with disabilities, and all have a greater capacity for emotional involvement and sustained commitment than do any agents of the state.

The state, it should be clear from this essay, exists as a *subsidium*, as a help or aid, to those pursuing their own human flourishing, one part of which includes carrying out of local and personal responsibilities. Respect for the principle of subsidiarity is thus a requirement of an adequate response on the part of the state to the needs of the disabled, as to the needs of all other persons. The state should not be in the business of taking over the care of the disabled—only the business of helping them, as possible, and their caregivers, to live their own particular lives.[9]

A second general point is that the needs that govern the justification for political authority are multiple: defense against outsiders, for example, and against internal criminality, as well as coordination of ways of life, partly by provision of infrastructural necessities such as roads and currency. All these activities require resources, which are finite, and whose source is, ultimately, the citizens of the state. This creates inevitable limits to what the state can do, and a need for judgments of prudence in deciding what is possible. It is simply not the case, given the inevitable finitude and frequent scarcity of resources, that the state can provide for every need, and often some needs which could in theory be better satisfied will in fact not be. Such limitations are not matters of injustice.

Moreover, in determining what will be done, what money spent, and what resources consumed by and for whom, the state must be fair as regards all its possible beneficiaries. The correlative to not viewing the disabled as a special interest group, but as on equal foundation with all as regards the justification for political authority, is, it seems, that their needs must be balanced against the needs of others which, while perhaps not as extensive, are nevertheless among the needs for which that authority exists. Consider this in the context of education: it is incumbent upon the state to provide resources by which those with special needs may be aided, but the needs of non-disabled children (and also exceptionally gifted children) must also be met. Finding the correct balance here is a matter of political prudence; there is no way of privileging one set of needs over the other here so as to arrive at a neat solution for the problem of distributing resources.

Social justice for the disabled is thus a rigorous requirement for political authority, though it is not the only, or an inevitably overriding commitment. But it is a commitment, I have argued here, whose grounding and nature can best be understood from a natural law, and not a classically liberal, perspective.

Notes

1. Nussbaum, too, notes a kind of inconsistency in Kittay on the matter of allegiance to liberalism (Nussbaum, 2002, p. 195).
2. The approach articulated in this essay, both ethically and politically, owes much to the work of the New Natural Law theorists. See especially Finnis (1980, 1998); Grisez, Boyle, and Finnis (1987); Grisez (1993). See also St. Thomas Aquinas (1988).
3. Cf. Aquinas: "If it is natural that man live in society with his fellows, it is necessary that there be some power in men by which that group be ruled" (Aquinas, 1988, p. 264).
4. For extended discussion of the case of patients in PVS, and the ways in which they can be benefited and harmed, see the essays in Tollefsen (2008), especially Fisher (2008) and Degnan (2008).
5. I do not, however, want to overstate the asymmetry. Many of the disabled, perhaps especially those who become so in the course of their lives, must meet moral challenges for which they may not have been adequately prepared. The soldier who loses a limb in combat can meet his new disability with grace and humor, or with bitterness and resentment. This too is a moral burden that should, to the extent possible, be shared by others.
6. For a sustained defense of these claims, see George and Tollefsen (2008).
7. Cf. Alasdair MacIntyre, in *Dependent Rational Animals*:

   > What I am trying to envisage then is a form of political society in which it is taken for granted that disability and dependence on others are something that all of us experience at certain times in our lives and this to unpredictable degrees, and that consequently our interest in how the needs of the disabled are adequately voiced and met is *not a special interest*, the interest of one particular group rather than of others, but rather the interest of the whole political society, an interest that is integral to their conception of their common good (Macintyre, 1999, p. 130, emphasis added).

8. The natural law standpoint further enables resolution of a tension seemingly intrinsic to the attempt to embrace both the needs of the disabled and the demands of liberalism, for the natural law standpoint urges the protection of the lives of *all* dependent human beings, both born and unborn. There is surely a deep tension, as Reinders points out, between the push to "normalization" for the handicapped, and the liberal willingness to allow women to abort handicapped children, a tension never adequately addressed, to my mind by liberal advocates of the rights of the disabled. Consider, for example, Michelle Fine and Adrienne Asch's statement that the right to abortion "for any reason" and the rights of newborn disabled to treatment are "separate rights." The crucial premise in arguing for this is that the fetus, unlike the newborn, exists within the womb, and "depends on the mother for sustenance and nourishment" (Asch & Fine, 1988, p. 302).

   But this argument uses as a premise the very claims concerning universal dependence that critics of liberalism's ability to address the problems of the disabled have focused on. The unborn are just like all of us (and all of us were once unborn) in their inevitable and radical dependence during at least some part of the lives; yet the state exists to meet those needs that we are inadequate to meet on our own. The state should protect the unborn for *exactly* the same reasons that it should protect disabled newborns, children, adults, and the elderly. Failure to do so clearly jeopardizes the disabled in two ways, both by encouraging their selective elimination, and by sustaining a legal situation in which some, the unborn, are discriminated against because their profound dependence is viewed as giving weight to the concerns of others against them. An adequate political approach to the needs of the disabled cannot abstract from the problem of abortion; but neither can it solve it only selectively, by addressing only the issue of abortion of the disabled. For this would be to reestablish again the same form of unfairness that those aborted, neglected, mistreated or abandoned because of disability suffer; that is, it would single out a class, in this case the non-disabled unborn, and allow action against them in virtue of their profound dependence. There can be no stability for the right reasons in a political landscape that protects the disabled but denies these protections to the unborn.

9. For a paradigmatic statement of the Principle of Subsidiarity, see Pius XI (1931, no. 79).

References

Asch, A., & Fine, M. (1988). Shared dreams: A left perspective on disability rights and reproductive rights. In M. Fine & A. Asch (Eds.), *Women with disabilities: Essays in psychology, culture, and politics* (pp. 297–305). Philadelphia: Temple University Press.

Degnan, M. (2008). Are we morally obliged to feed PVS patients until natural death? In C. Tollefsen (Ed.), *Artificial nutrition and hydration: The New Catholic Debate* (pp. 39–61). Dordrecht, The Netherlands: Springer.

Finnis, J. (1980). *Natural law and natural rights*. Oxford: Clarendon Press.

Finnis, J. (1998). *Aquinas: Moral, political, and legal theory*. Oxford: Oxford University Press.

Fisher, A. (2008).Why do unresponsive patients still matter? In C. Tollefsen (Ed.), *Artificial nutrition and hydration: The New Catholic Debate* (pp. 3–37). Dordrecht, The Netherlands: Springer.

George, R. P. (1993). *Making men moral*. Oxford: Oxford University Press.

George, R. P., & Tollefsen, C. (2008). *Embryo: A defense of human life*. New York: Doubleday.

Grisez, G. (1993). *The way of The Lord Jesus Christ: Living a Christian life* (Vol. 2). Quincy, IL: Franciscan Press.

Grisez, G., Boyle, J., & Finnis, J. (1987). Practical principles, moral truth, and ultimate ends. *American Journal of Jurisprudence, 32*, 99–151.

Kant, I. (1970). On the common saying: This may be true in theory, but it does not apply in practice. In H. Reiss (Ed.), *Kant's political writings* (pp. 61–92). New York: Cambridge University Press.

Kittay, E. F. (1999). *Love's labor: Essays on women, equality, and dependency*. New York: Routledge.

Locke, J. (1986). *Second treatise on civil government*. New York: Prometheus.

MacIntyre, A. (1999). *Dependent rational animals: Why human beings need the virtues*. Chicago: Open Court.

Nussbaum, M. C. (2000). *Women and human development: The capabilities approach (The Seeley lectures)*. Cambridge, MA: Cambridge University Press.

Nussbaum, M. C. (2002). The future of feminist liberalism. In E. F. Kittay & E. K. Feder (Eds.), *The subject of care: Feminist perspectives on dependency* (pp. 186–214). Lanham, MD: Rowman and Littlefield.

Nussbaum, M. C. (2006). *Frontiers of justice: Disability, nationality, species membership*. Cambridge, MA: Belknap Press.

Pius, XI (1931). *Quadragesimo Anno* [On-line.]. Available at http://www.vatican.va/holy_father/pius_xi/encyclicals/documents/hf_p-xi_enc_19310515_quadragesimo-anno_en.html

Rawls, J. (1971). *A theory of justice*. Cambridge, MA: Harvard University Press.

Rawls, J. (1980). Kantian constructivism in moral theory. *Journal of Philosophy, 77*, 515–571.

Rawls, J. (1996). *Political liberalism.* New York: Columbia University Press.

Reinders, H. R. (2000). *The future of the disabled in liberal society: An ethical analysis.* Notre Dame: University of Notre Dame Press.

Rousseau, J. J. (1997). *The social contract and other later political writings.* Cambridge: Cambridge University Press.

Silvers, A. (1998). Formal justice. In A. Silvers, D. Wasserman, & M. Mahowald (Eds.), *Disability, difference, discrimination: Perspectives on justice in bioethics and public policy* (pp. 13–146). Lanham, MD: Rowman and Littlefield.

St. Thomas Aquinas. (1988). *On law, morality, and politics* (W. P. Baumgarth & R. J. Regan, Eds.). Indianapolis: Hackett.

Tollefsen, C. (Ed.). (2008). *Artificial nutrition and hydration: The New Catholic Debate.* Dordrecht, The Netherlands: Springer.

Chapter 14
The Unfair and the Unfortunate: Some Brief Critical Reflections on Secular Moral Claim Rights for the Disabled

H. Tristram Engelhardt, Jr.

14.1 Facing Tragedy: Not All that is Unfortunate is Unfair

Disease, early death, disability, and suffering mark the human condition. The hopes and aspirations of humans are often tragically and indeed inevitably foreshortened by illness and disability. We are finite mortal beings. Medicine is shaped by and directed to the amelioration of the tragic limitations and often uncontrollable suffering that burden much of human life. The response to the disease, suffering, and disability of others should be characterized by care and love. Failures of appropriate response to the needs and suffering of others are subject to moral judgments of being uncharitable, unkind, unfeeling, and unloving. However, it is far from self-evident that the domain of the unfortunate that encompasses disease, disability, and suffering can with conclusive secular rational justification be translated into claims of unfairness. That bad things happen and are an element of the human condition is surely unfortunate. The question is the extent to which this feature of human tragedy can be understood as being unfair in the sense of being a ground for claims based on considerations of justice that require others to provide care, support, and accommodation with such compelling character that it can warrant state enforcement. For the purposes of this article, claims to care, support, and accommodation grounded in considerations of justice are taken to have a force, if they are warranted, that can require the compliance of others, including the use of their property and their compliance to various policies established for the benefit of the disabled.

The problem is whether a canonical moral warrant can through secular moral reflection be secured for such governmental interventions and the claims they establish. An examination of the possibility of securing a general secular moral warrant for the establishment of claim rights for the disabled to care, support, and accommodation not only brings into question the general secular moral warrant for such claim rights, but it brings into question as well the secular moral authority of the more than minimal state. This brief set of reflections is focused on attempts without

H. Tristram Engelhardt, Jr. (✉)
Department of Philosophy, Rice University, Houston, TX, USA
e-mail: htengelh@rice.edu

D.C. Ralston, J. Ho (eds.), *Philosophical Reflections on Disability*, Philosophy and Medicine 104, DOI 10.1007/978-90-481-2477-0_14,
© Springer Science+Business Media B.V. 2010

sufficient argument and reflection to translate the unfortunate into the unfair with regard to the needs of the disabled, yet it discloses beyond this issue a major crisis of moral authority in the post-Christian West. If the warrants for such claim rights cannot be secured, we will be forced foundationally to re-consider the moral force of law and public policy in these and other areas.

14.2 Justice-Based Claims to Support, Care, and Accommodations for the Disabled: Some Skeptical Reflections

We approach reality within a thick web of moral, empirical, and metaphysical taken-for-granted expectations and commitments. These expectations and commitments direct us in sorting information from noise in our empirical, moral, and metaphysical assessments of, and judgments regarding, reality. The knowledge and action of finite beings is contextual and historically conditioned. This is surely the case when one makes judgments regarding disabilities and advances claims on behalf of the disabled. Diagnoses of diseases, disabilities, and defects are nested within framing scientific, moral, and metaphysical assumptions set within socio-historically structured environments. Diagnoses are articulated in terms of expected ranges of accepted function, form, freedom from distress, and lifespan (Engelhardt, 1996, Chap. 4). To have a disability is, for example, not to have an expected ability. Diagnoses of diseases, disabilities, defects, and illnesses not only describe reality (e.g., "John Smith is deaf; he has complete hearing loss") but also serve as a means for evaluating reality in the sense of recognizing a person's abilities as falling short of some norm (e.g., as when holding "John Smith is deaf; he is disabled"). The norms at stake are non-moral norms, but they nonetheless support evaluations that lead to the characterization of certain circumstances as good or bad, better or worse (consider the non-moral judgment involved in holding that "this wine is the best of the three"). Diagnoses also indicate the advisability of seeking a particular remedy (e.g., "Perhaps a hearing aid would help"). Diagnoses also explain reality by identifying underlying conditions known to produce sickness, disease, and disability (e.g., through offering accounts of the etiology and pathogenesis of the disability).

Beyond the descriptive, evaluative, and explanatory functions of diagnoses, diagnoses have a performative function. They create social roles. Determinations of disability place the disabled within particular social roles that can be apprehended on analogy with the sick role. That is, the sick role and the social role of being disabled locate a person within a socially constructed context that relieves the person in that role of certain duties (e.g., military duty) and provides grounds for excusing from blame (e.g., "He was not ignoring you; he did not hear you because he is deaf"), while establishing social expectations (e.g., to seek treatment) and recognizing certain professions as in authority to make recommendations to the sick and disabled (e.g., physicians) (Parsons, 1951, 1958; Siegler & Osmond, 1973). In addition, a particular diagnosis can lead to welfare claims (e.g., the finding of a patient as totally and permanently disabled can qualify a patient for the provision of a certain level of benefits within an established insurance system), as well as payments for treatment (e.g., under an applicable insurance plan).

Diagnoses thus establish warrants for social interventions within a set of sociohistorically conditioned expectations that establish a rich web of social responses. All of this is placed within a social context supported by the assumption of various background norms, both moral and non-moral (e.g., regarding proper ranges of function), including socially and politically established expectations of charity and welfare that are held to be appropriate responses to disease or disability (Engelhardt, 1996, Chap. 4). In the fashioning of this social space for therapeutic and other social interventions, there is an intersection of medical and social models that provide accounts of the cause, pathogenesis, and treatment of disease and disabilities. Medical models provide accounts of the genetic, bacteriological, or traumatic etiology, the pathogenesis, the underlying pathology, and possible medical interventions bearing on disabilities. Social models provide accounts of the ways in which social attitudes towards disabilities as well as social responses to states of disability can exacerbate or ameliorate disabilities. There is a medical and social dimension to most concerns with disease and disability, ranging from diabetes to deafness. These concerns are often tied to alleged claim rights to care, support, and/or accommodation.

A cardinal question is thus confronted: is the fabric of claim rights advanced on behalf of the disabled purely a social construction (e.g., the creation of claims as through an insurance policy), or is it grounded in general obligations of justice that oblige persons as such? That is, is there something about moral rationality or the human condition that rationally obliges humans to recognize such claim rights to care, support, and/or accommodation as part of the recognition of disease and disability? A significant dimension of the debates concerning allocations of resources for health care reflect disagreements as to when and under what circumstances health care needs generate rights to health care, as well as against whom claims grounded in such rights could properly be directed. So, too, with issues of disability. Disease and disability are surely, *ceteris paribus*, unfortunate. The issue is whether they are unfair in a way that generates general secular moral claim rights against others who did not cause the disease or disability. Some of the essays in this volume appear to take for granted that those with disabilities have such claim rights grounded in considerations of justice against society, as well as against intermediate institutions and individuals, to provide care, to offer support, and to modify environments so that the disabilities of the disabled will be less encumbering. Here the language of differently abled is at times turned to the claim that society or others should change environments so as to make disabilities less disabling. One might consider the provision of access by those using wheelchairs to public places, as well as to privately owned businesses.

Critically to address the force of claims on behalf of the disabled to such social welfare grounded in accounts of justice and fairness is not to bring into question the view that it would be charitable, kind, and beneficent to provide such care, support, and accommodation for the disabled. Although it may be the case that claims supposedly grounded in considerations of justice tend to marginalize charitable action, thus undermining the nurturing of virtue in individuals, intermediate institutions, and society generally, the focus here will be on the general force of such claims of unfairness. The issue is whether there are claim rights to care, support, and accommodation for the disabled as a matter of justice, as well as regarding what level of costs and inconvenience entailed by responding to such claim rights would through their burdens defeat those claims. Given the limited scope of this brief article, it will be enough here to show that there are good grounds for recognizing that secular moral claims on behalf of the disabled are not self-evident, despite however well established they might be in positive law and popular culture. Beyond that, this essay will only be able to indicate the foundations of a profound crisis regarding moral vision and political authority in the secular, post-Christian West.

14.3 The Challenge to Claim Rights on Behalf of the Disabled

At the outset, in order to establish claim rights of the disabled to special care, support, and/or accommodation, one would need to show why it is obligatory to assume a moral point of view from which it will be judged not only unfortunate that persons of a particular class are disabled, but that this state of affairs is (1) unfair and (2) generative of claim rights against society as a whole or against particular groups of people (e.g., proprietors of private business establishments) for support, aid, or accommodation to their disablement in a way that morally should create enforceable obligations against possible caregivers that are not outweighed by the property rights, autonomy rights, or burden of those who would be obliged to provide care, support, and accommodation. The difficulty is that we do not share a common morality, if one means by a common morality a common understanding of when it is forbidden, licit, or obligatory to have sex, transfer property, or take human life. Instead, our cultures are marked by controversies over these foundational issues, because the participants in the controversies are moved by incompatible moral visions framed in terms of different rankings of primary social goods, as well as of cardinal right- and wrong-making conditions.

The issue of the obligation to make accommodations for the disabled is of particular interest, because the scope and force of a disability will usually vary, depending on the environment in which the disabled find themselves. In particular, if funds were invested in changing the physical environment of the disabled, many of the disabilities of the disabled would not be as severe, pressing, or limiting. Therefore, it is in general good, all things being equal, to alter the environment so as to make the disabilities of the disabled less encumbering. However, all else is rather rarely equal. When raising the issue of claim rights to have the environment altered, one confronts the root question as to why anyone would be obliged from considerations of justice to provide care, offer support, or change the environment so as to ameliorate the circumstances of the disabled. That is, even if it would be good, *ceteris paribus*, to act to set such limitations aside, one must show why it would be obligatory to do so. In doing so, one must distinguish between what is good in the sense of supererogatory to do from what is obligatory to do especially as a matter of justice. There is a sense to the claim that "yes, I ought to do X, which is a good thing to do, but you have no right to compel me to do this good act X." The goodness and obligations of charity are easier to establish than claims grounded in a view of justice, in particular when it would lead to the compulsion associated with public policy and law. Because the latter would limit the property rights and constrain the freedom of action of those against whom the claim rights are directed, further argument is needed. Claim rights to care, support, and/or accommodation that conflict with property rights to the enjoyment, use, and design of one's own resources bear a greater burden of proof because they must establish their authority to warrant such interventions in the absence of the consent of those invoked.

14.4 Justice in the Face of Moral Pluralism: The Deflation of the Self-Evidence of Claim Rights on Behalf of the Disabled

A cardinal difficulty facing the project of defending a generally compelling account of justice that would support the view that the disabled have a *prima facie* claim to care, support, and accommodation for the disabled is that we do not share one vision of justice or a common understanding of how needs justify rights, much less a common morality. John Rawls' (1921–2002) early work provides an example of the arbitrary character of all such accounts of justice. Rawls' account of justice as fairness builds its initial plausibility around his invitation to envisage how one would allocate resources and social status, not knowing from the hypothetical perspective of the original position the place one would have in the society one is designing. The idea is that in approaching the structure of society in this fashion, one would as a hypothetical contractor design a fair society, one against which no one, including oneself, could rationally protest, no matter what resources or social status one possessed. That is, persons are asked to imagine entering into a hypothetical social contract through which they will allocate resources and social positions, not knowing their future status in that society, so that whatever position they received is that to which they would have agreed within the hypothetical standpoint of the original position (Rawls, 1971). In his appeal to the way hypothetical contractors would structure social commitments, Rawls builds in a central place for equal opportunity. The idea is that the contractors would act to ensure themselves against being disadvantaged and therefore affirm a principle of equality of opportunity.

In order for this strategy to warrant a particular set of social responses, the hypothetical contractors invoked for this ideal moral perspective must be fitted out with a particular thin theory of the good (a particular lexical ordering of liberty, equality, and prosperity) and a particular view of risk-aversiveness (i.e., that one would not be willing to risk great disadvantage in the pursuit of possibly even greater advantage), so as to produce the conclusions desired regarding welfare claims. A different thin theory of the good, a different sense of appropriate risk-aversiveness, will authorize a different sense of justice as fairness. For example, a thin theory of the good that gives priority to security and prosperity over equality and liberty will produce justifications not for a social democracy, but for a one-party dictatorial capitalism. That is, given different basic assumptions, the hypothetical contractors will endorse quite different political and social structures with differing views as to fair equality of opportunity. Without any background value assumptions, no particular moral vision or particular public policy will be endorsed. What one can justify depends on what one feeds into the initial background commitments. Appeals to a reflective equilibrium serve as means for adjusting conclusions and principles so as to secure and flesh out a warrant for the moral lifeworld one wishes to offer. One is thus able to create a justification for the particular fabric of social policies one favors by feeding the right content into the background thin theory of the good and view of moral rationality.

The Rawlsian approach to the justification of his particular account of justice or fairness is thus arbitrary, in the sense that it relies on the selection of a particular set of background assumptions, including a particular thin theory of the good, as well as of proper risk-aversiveness, for which no conclusive sound rational argument can be given (Engelhardt, 1996, chaps. 1–3, 5). This state of affairs can be underscored by noting that the Rawlsian approach as originally structured leads to endorsing outcomes that many would take to be a *reductio* of Rawls' position. For example, if one is to arrange the distribution of social status and of resources so as to be to the benefit of the least-well-off class, the question then arises as to what class constitutes the least-well-off class. If the least-well-off class is the class of those who die before the age of 18, and if one endorses a Rawlsian principle of fair equality of opportunity, and if health care is regarded as governed by this last principle, then resources would be obligated first and foremost to pediatric interventions before any resources would be available to be allocated to the disabled over 18 years of age.

The point is that an account of claim rights for the disabled is dependent on endorsing a particular ranking of cardinal human goods. The arbitrary content of Rawls' thin theory of the good is underscored when one recognizes the intellectual burden of justifying the requirement of affirming his thin theory of the good and his particular view of appropriate risk-aversiveness that characterizes the perspective of Rawls' original position. In addition, a cardinal assumption of the position Rawls advances is that one must consider talents and resources as unowned until justified by a practice of ownership grounded in and warranted by something like the principles Rawls seeks to establish through his original position. This account of the original position, along with its particular understandings of a proper thin theory of the good and its account of risk-aversiveness, predestines the kinds of conclusions that Rawls seeks to secure. However, the question then is why ought one to grant these background premises.

The problem of justification does not simply attend Rawls's account. The problem besets all attempts to provide a general, rationally justifiable, content-full, secular account of justice or fairness, and it cannot simply be wished away. All views of just and fair conduct, as well as understandings with regard to what it means to realize fair equality of opportunity require in the background a particular understanding of moral rationality, a particular moral sense, a particular thin theory of the good, a particular ranking of primary human goods, and/or a particular understanding of how to compare different forms of preference satisfaction. The problem is that without a background canonical moral perspective, which would be the secular equivalent of the perspective of God, one cannot know how to compare rational versus passionate preferences, corrected versus uncorrected preferences. Nor can one know with the warrant of a conclusive sound rational argument the secular equivalent of God's discount rate for preference satisfaction over time. Set within the sphere of the finite and the immanent, one cannot identify the canonical content that should inform the moral sense, the thin theory of the good, the proper ranking of goods, or the account of preferences that should be normative, without begging the question, arguing in a circle, or engaging in an infinite regress. This is the case because particular understandings of moral rationality, of the appropriate moral sense, and/or of the appropriate thin theory of the good depend for their general secular rational justification on particular background moral premises and rules of moral inference, which are themselves matters of controversy. Moral diversity, the fact of moral pluralism, undermines the self-evident character often attributed to claim rights for care, support, and accommodation (Engelhardt, 2006).

14.5 From Justice to the Creation of Social Insurance

Given a plurality of visions and accounts of what is fair and unfair, indeed in the face of a plurality of moralities supported by a plurality of possible background moral premises and rules of moral inference among which one is not able to choose definitively on the basis of sound rational argument without begging the question, arguing in a circle, or engaging in an infinite regress, one is left, if one is committed to advancing claim rights on behalf of the disabled, with simply creating particular social responses to the plight of the disabled, as occurs within social secular democracies. Particular social responses to the circumstances of the unfortunate become particular political creations, not moral discoveries. From the general perspective of a secular democracy, social transformations of the unfortunate into the unfair (e.g., by creating welfare rights and rights to accommodation for the disabled) cannot be secured or anointed by a canonical account of justice. Such claim rights to care, support, and accommodation must from a general secular moral perspective be understood to be the product of political negotiation and compromises made in the course of democratic processes, or in the case of one-party dictatorial regimes, the establishment of a particular policy by a particular ruling elite. Entitlements are in the course of such happenings created, not discovered. They lose any claim to general secular moral standing or authority. They are lodged in politically fashioned webs of social insurance whose authority and content cannot be canonically blessed by a secular, canonical, content-full view of fairness and justice. The question then remains, can any entitlement claims on behalf of the disabled have any moral and not simply political significance or standing?

If there are no sound rational arguments to establish canonical moral claims on behalf of the disabled, then one might still hope that the procedures of democracies would be able to convey moral authority to their creation of social insurance programs that could support entitlements for the disabled. In such circumstances, one might hope for a derivative moral authority that could follow from a morally authoritative process. However, such moral authority will prove difficult to secure, for the same reason that it is impossible by sound rational argument to establish a canonical account of how to regard unfortunate circumstances as unfair circumstances. This is particularly difficult as one steps beyond the bare authorization of the minimal state to seeking moral authorization for social entitlements from the more-than-minimal state (Engelhardt, 1996, Chap. 5). If the more-than-minimal state has no secular moral authority to establish extensive welfare rights, then supposed entitlements of the disabled to care, support, and accommodation will merely be the result of strategic political compromises that establish and sustain a political *modus vivendi* in which a sufficient portion of the society is willing to acquiesce so as to ensure political stability, including the rule of law. In such circumstances, entitlements for the disabled will not enjoy a secular moral authority. They will simply be outcomes that it will usually be prudent to accept. One will not just confront a plurality of moralities and bioethics (Engelhardt, 2006), but one will in addition be confronted with a crisis in the moral grounding of political authority.

This loss of general moral authority for social programs is a function of the difficulty of providing a secular surrogate for the religious justification of secular state authority, which religious justification supplied a source of authority that up until recently played a key background role in the Western political world-view. Christians, as St. Paul warns, are to "let every person be subject to the governing authorities; for there is no authority except from God, and those authorities that exist have been instituted by God. Therefore whoever resists authority resists what God has appointed, and those who resist will incur judgment" (Romans 13:1-2). Absent the recognition of this endorsement of the divine right of kings and/or democracies, one is left with a challenge to which secular moral reflection does not appear equal. The moral authority of the fabric of the more-than-minimal state, along with its welfare policies, is not just radically deflated (i.e., reduced to being nothing more than that which is supportable within the minimal state), but that which is not deflated is radically recast in its authority so that it rests on a *modus vivendi* to which it can be plausibly prudent to acquiesce, but for which there is no general secular moral authority. Much work will need to be invested in exploring how we should understand the status of the purported moral claim rights advanced on behalf of the disabled supported by secular understandings of justice, such as those that are established by secular democracies. The difficulties are radical. Even the secular authority of such democracies is brought into question in the face of the in principle irresolvable moral pluralism that marks the human condition in a post-Christian world. The point is not to deny moral pluralism within the Christian moral and political understandings of the past. However these understandings had metaphysical commitments that impelled them to proceed as if there were a canonical, moral, and epistemological perspective in terms of which moral and metaphysical controversies could be authoritatively resolved. There are no longer grounds within the ambit of the contemporary post-Christian culture of the West to support such an assumption. The questions raised by the pursuit of claim rights for the disabled to care, support, and accommodation disclose and underlie a significant crisis in the secular moral authority of the state.

References

Engelhardt, H. T., Jr. (1996). *The foundations of bioethics* (2nd ed). New York: Oxford University Press.

Engelhardt, H. T., Jr. (Ed.). (2006). *Global bioethics: The collapse of consensus*. Salem, MA: M&M Scrivener Press.

Parsons, T. (1951). *The social system*. New York: Free Press.

Parsons, T. (1958). Definitions of health and illness in the light of American values and social structure. In E. G. Jaco (Ed.), *Patients, physicians and illness* (pp. 156–87). Glencoe, IL: Free Press.

Rawls, J. (1971). *A theory of justice*. Cambridge, MA: Harvard University Press.

Siegler, M., & Osmond, H. (1973). The "sick role" revisited. *Hastings Center Studies, 1*, 41–58.

Part IV
Personal Voices

Chapter 15
Neither Victims Nor Heroes:
Reflections from a Polio Person

Jean Bethke Elshtain

Polio is suddenly a hot topic. This should not surprise us, given our shared cultural urgency to tell our stories. As the World War II generation began to fade from cultural and living memory, those recognizing this fact rushed to generate hundreds of films, articles, books, television productions and monuments to "the greatest generation." Now that we "polios" have entered "mature" age and have experienced the onset (for tens of thousands of us) of a "post-polio syndrome" that has thrown us for a loop physically and taken us aback psychologically, this creates a "perfect storm" of the sort that inspires; jogs often-painful and unpleasant memories; and compels people to "go public." Nearly all the polio books published over the last decade have been written by polio people themselves or by others who have a friend, family member, or acquaintance whose polio had a direct effect on them.

No one wants to go to the grave unheard, with his or her story untold. There's that. There is also the fact of how society "names" us and how we think of ourselves. If, at one point, being a "polio victim" was a stigma, a sign of infirmity, something frightening to behold, now we are told that we are not only survivors but "polio heroes" who have persevered where lesser souls might have quaked and quivered. Surely, however, post-polios are neither victims nor heroes—just a lot of folks (the estimates are that around 225,000 of the 600,000 or so living polio survivors have been hit with post-polio syndrome) who got a rotten deal once, and then a second time—and who are, like everyone else, trying to make the best of the hand they have been dealt. It isn't always easy, of course—but, then, life never is.

In this essay I will explore the phenomenology of victimization and "heroization" (with apologies for the clunky word), arguing that the story of polio, whether on the societal or individual level, has lurched back and forth between these two conceptual poles and types of accounts—victim/hero—which accounts ill serve concrete, living human beings whose bodies were and are marked by an unusually cruel disease that most often laid low the young.[1]

J.B. Elshtain (✉)
Divinity School, University of Chicago, Chicago, Illinois, USA
e-mail: jbelshta@uchicago.edu

D.C. Ralston, J. Ho (eds.), *Philosophical Reflections on Disability*, Philosophy and
Medicine 104, DOI 10.1007/978-90-481-2477-0_15,
© Springer Science+Business Media B.V. 2010

15.1 Polios as Victims

In what counted as our hometown newspaper, the *Fort Collins Coloradoan* (my little village, Timnath, Colorado, population 185, didn't rate a newspaper of its own), a photo of a plucky polio kid, staring determinedly at the camera, the hint of a smile on her face, posed with crutches under each arm and her left leg encased in a heavy steel and leather brace, appeared under the caption: "That kind of spirit polio can't beat!" The plucky kid was me, age 11, and I had been crowned a polio hero, having "defeated" a crippling disease. The evidence? I was on my feet. I had "learned to walk again"—the lodestar, the touchstone, for all of us. Being upright, neither bedridden nor "wheelchair bound" was the brass ring, the Olympic medal. I certainly experienced it as a great victory.

For all of us polio kids, there were choices—not, for the most part, voiced but clear nonetheless. One was "giving up," capitulating to the illness and to one's new-found helplessness. A second was accommodation: "O.K., you got me, but let's meet each other half way." A third was "triumph-over": polio gets none of me; I refuse to be seen as a victim, I'm "normal" and I'll prove I'm such. Period. "Giving up" meant one was an object of pity—also contempt for not having tried. The second option seemed a bit like settling for a B when, if you worked a bit harder, you might have received an A. Not acceptable. Instead, for those of us who opted for "triumph over," it was a war and the endpoint was victory, not a negotiated settlement. This latter ideal meant that the victim/hero dichotomy was securely in place: it was one or the other, or both simultaneously.[2] There was no middle ground and scant room for ambiguity.

The iconography within which each polio person was lodged as part of an inescapable cultural surround was drenched in bathos: scenes of polio kids lined up in wheelchairs, or submitting stoically to painful physical therapy, or reveling in the visit of a celebrity to the polio ward while little polio victims beamed at the attention (in my case, that of television cowboy Hopalong Cassidy). Appeals for donations from the March of Dimes invariably featured a plucky kid, standing upright, facing the world braced and crutched (if there is such a word). Many disability rights groups, looking back, are incensed at these appeals, finding them demeaning, turning the polio kid into a crippled advertisement for the "conformism" of the Eisenhower era, or some such ideological trope. One can understand this up to a point, I suppose. I'm sure many adult polios at the time mocked the polio poster child icon.

I confess I rather wanted to be that child. I was plucky! I was determined! I could look appealing! Slightly embarrassing in retrospect, to be sure, but entirely understandable. Getting on your feet meant that you would not be an object of pity, although the victimization backdrop never disappeared. One saw it in the eyes of folks who greeted your parents after church, or in other social settings, as they voiced their sorrow at your parents' suffering (they, too, were victims, being parents of a polio child). These same well-meaning folks often gazed at you with doleful eyes. Although you might, by then, be up and about with the aid of brace and crutches, you were a victim all the same. Those who were Christians tried to place the polio epidemic within a theodicy of some sort. Surely God had *something* in mind, some purpose. None were so cruel as to suggest one's victimization was a specific punishment for an unspecified sin, although some may have thought this. Most, however, surely echoed Abraham Lincoln: God's purposes are not our own. We cannot discern the Divine will.

How did we victims/heroes respond to the theodicy question? (For we forget to our peril—if our goal is understanding and explanation—the pervasive fact of faith in the lives of the overwhelming majority of American families in the 1950s, the years of the last great polio epidemics.) There is little doubt in my mind that, for most polios, the recitation of the Scriptural assurance that all things work out to the good for those who love God, could leave a bitter taste on the tongue. Still, if one was a Christian polio child, one was hard pressed to make sense of one's sudden paralysis in light of the Creation account: the earth and all that is in it was brought into being by a loving God who "so loved the world" that He gave His only begotten Son to die for one's sins and those of all humankind.

In our Sunday school we sang: "Jesus loves the little children/All the children of the world/Be they yellow, black, or white/They are precious in his sight/Jesus loves the little children of the world." So if Jesus loves me and I am part of a good creation brought into being by a loving God, then I have not been singled out for arbitrary and cruel punishment. There is a purpose here nonetheless, I decided, and the purpose is that God is putting me to the test, like the early martyrs. They loved God and He loved them. If they gave their lives, I can certainly give my best effort and honor God thereby. Too, the martyrs were not whiners. Did they shiver and shake in the Roman arena and cry: "Boo hoo. Why me?" Certainly not.

God is omniscient. He knows this will not defeat me. That does *not* mean, however, that my own agency makes no difference. It makes *all* the difference, as God "helps those who help themselves," a saying drummed into my consciousness throughout the almost 10 years of life I had lived before I was "stricken," in the language of the day, with polio. So I was on a quest. My purpose was to pass God's test, to demonstrate my mettle, and to prepare myself for any later test that God, in His wisdom, might put me through. Yes, I know, this is pretty simple theology. But it served the 10 year old I was rather well.

For it could not fail to be the case that those of us who were able to locate our own travail within a powerful teleology, a narrative with a purposive direction that enlisted our efforts to reach its culmination, did "better" than those who sank into bitterness, convinced that God, or someone, or something, had punished them, that they had become akin to criminals with no idea of what were the crimes that warranted the punishment they were suffering. If all of this added up to "normalization," in today's lingo, so be it. Better that than to languish in bathos, on the victimization side of the victim/hero coin.

A few additional thoughts on polios as victims: it was impossible for any polio "victim" and his or her family to escape the victim designation altogether, even if the child located herself on the hero side of the ledger. For this was the preferred designation of the time, and there was nothing subtle about it. "Victim," however, is an inherently tricky word. We speak, indifferently, of murder victims and earthquake victims, but there is a huge difference between these two categories. Earthquake victims are utterly random. Seismic shocks do not single out the "Smith Family on River Road" to terrorize and torment. A murder victim may be random, but is more often than not singled out for death by a murderer. Direct intentionality is involved.

Polio is surely more on the earthquake side of the continuum. And yet. . .many there were, including polio "victims" and their families, who bore a heavy load of debilitating guilt. "If only I hadn't let her go swimming on Labor Day. . ."; "If only I had stopped him from heading out with his pals for the day at the circus. Circuses are filthy. . .". From the perspective of the polio kid, one might hear: "I should have done what Mom told me and been more careful not to swallow water in the swimming pool. . ."; "I shouldn't have gone to Susie's slumber party. I wasn't supposed to get overtired. . .". Plaints of this sort were common and altogether intelligible as people struggled to infuse what was happening to them with purpose and agency: I might have avoided this somehow. For the most part, an avowal of a measure of agency is a good thing.[3]

Those who understood that they were random victims of a horrible disease engaged their agency in acting out the *telos* of recovery, even victory-over (with God's help and that of family and friends). Those who carried a burden of guilt found their agency glued to the spot, so to speak, against the onslaught of symptoms and the dreaded diagnosis (most often confirmed by an extraordinarily painful spinal tap, in which a menacingly long needle pierced one's body in order to test the spinal fluid) and the verdict—thumb's up or thumb's down—was awaited with fear and trembling.[4] Lives could be, and were, forever overshadowed by a burden of guilt as a result of a polio diagnosis. My mother and father never got over my polio. For my mother, especially, it was a catastrophe. It blighted her life. She was certain there were preventive measures she should have taken—this from a woman who, in those days, scrubbed the floors twice a day!

When it turned out that an excess of cleanliness might itself be to blame— children in hyper-clean North America were never exposed to the poliomyelitis virus as children and, therefore, had generated no antibodies against it—it shook parents up all over again. This made the guilt even more devastating as the measures my mother took to protect her children from the disease may actually have contributed to one of her children being felled by it. Not only did I not do enough, a parent lamented, but what I did harmed my child.

To sum up: polio people who got stuck in victimization burnished daily the victim-guilt complex of their parents or spouses, thereby inaugurating, at first no doubt inadvertently, a cycle of victim-guilt-recrimination. What do I mean by "inadvertently"—for this suggests that, at one point, the cycle is intentionally maintained? Precisely this: there are certain advantages that flow from being a victim, especially if you are ministered to by a guilt-ridden caregiver who, by definition, can never do enough for you. It was clear to my 10-year-old eyes and ears that a number of polio kids in Ward F, Children's Hospital, Denver, Colorado, had taken that course—with some even going so far as to embrace infantilization, a second babyhood: "I'm helpless. It's your job now to do everything for me."

I do not mean in any way to diminish the arduous travail of those polio rendered almost completely helpless, e.g., those in iron lungs slated to spend their lives in the infernal yet life-saving machines. Even those of us who knowingly wanted to lift the burden of guilt from our parents usually failed, in whole or in part, in that effort—such is the potency of the victim-guilt-self-recrimination narrative.

15.2 Polios as Heroes

The reader is by now alert to the direction these reflections will take, given that "hero" is the opposite side of the "victim" coin and, further, that heroization is *always* encased within a frame of victimization—and vice versa—when one considers such mordant matters as polio epidemics, school shootings, or a range of other horribles, both natural and the result of human artifice, or some combination of the two. The hero triumphs over a catastrophic event, condition, or disease—stares it down and overcomes. Many of the important and vital post-polio support groups that now dot the landscape of American civil society, label post-polios as "heroes."

Again, in no way do I wish to criticize these efforts: they have made and are making an enormous contribution to the well-being of post-polio persons by drawing them out of isolation, helping to assuage their initial confusion, often followed by anger and depression, at new weaknesses and the loss of powers they had arduously attained in their "triumph-over" the disease.

Learning to live with loss and debilitation, perhaps a return to crutches, braces, or—God forbid (for many)—wheelchairs and other euphemistically-labeled "assistive devices" is, for many, like living out a nightmare. The nightmare is made all the more disagreeable given the pain and fear that is an invariable feature of muscular loss and spasms, breathing difficulties (for many), and all the rest. For some post-polios, the prospect of more surgery and lengthy hospitalizations, the precipitous end to one's active life in a career or as a hands-on homemaker or grandparent, presages serious psychological distress. None of this is easy to deal with, especially if you have located yourself on the heroic "triumph-over" end of the continuum for thirty years or more.

The good folks active in the post-polio movement think of "hero" not so much as moral supererogation but as a quotidian achievement: one is a hero simply for having survived the damnable disease.[5] But to those of us who pushed as hard as we could, gritting our teeth through the pain of physical therapy, ours was a victory that went way beyond survival and "learning to live with polio."[6] We had stared death—always a possible denouement with a polio diagnosis—in the face, and we were, by God, going to clobber polio itself, no TKO but a knockout in the 10th round, if it took that long. Now we were losing, for "it" had been lying in wait all these years, primed to ambush us. Enough already! Not so, as it turns out. Polio, the gift that keeps on giving!

So where does that leave post-polio heroes? In a strange cultural limbo, I suspect, *unless* one gracefully accedes to newfound victimization. Let me explain. Even as the hero triumphs, the victim recedes—assuming these are the dominant categories framing a cultural narrative. It follows that as the victim grows in strength, so to speak, the hero weakens. Post-polios are like Superman stuck in a sealed room filled with kryptonite. We are doomed: the kryptonite might not "get" us right away, but get us it will. This no doubt overdramatizes—but the heroic narrative did turn us into super-heroes in our own eyes and that of others, at least for a time, until the "triumph over" came to be regarded as "normal" and went unremarked. And that is just the way one wanted it. Although, for example, I could never run again (except in my dreams), I walked all over the place. When I first detected some signs of weakening in the late 1970s and early 1980s, I embarked on an exercise regimen, with walking the "exercise"—the very worst thing I could have done, as it turns out. Dealing with post-polio is a counter-intuitive task. What you think will forestall more weakening—namely, exercise, building up muscles, etc.—actually does damage, because you are generating more wear and tear on already worn-out ganglia that have borne a double or triple burden since the moment the onset of polio destroyed ganglia "originally" assigned to carry out a particular task, which task ancillary ganglia then had to take up. This is a rough-and-ready way of putting it, but perhaps the reader takes the point.

It took some time to sort this out. In the meantime, I remember thinking "by God, 'it' isn't going to get me this time around, any more than it did the first time." I would continue to pass. For decades, polio persons, in line with the hero narrative, asked for no "special favors." If the school had no elevator, no matter—one mastered the steps. If a college physical education requirement involved swimming a certain number of laps, one did it—there were be no special norm for me. But the inevitable could be forestalled only for so long, especially given the lamentable fact that one's efforts to fight growing weakness were exacerbating the symptoms one hoped to abate or to eradicate altogether.

When one was body-slammed with the reality of post-polio, one then had to face doctors—if indeed one could find a physician knowledgeable about polio who admitted to the reality of post-polio syndrome, given the accumulation of clinical data about the syndrome. The advice meted out, much of the time, indicated you were now to live a life that was more or less a reversal of the life you had been living if you were behaving like a hero: active, achieving, no concession to one's status as a childhood polio victim, for one had both grown up and grown out of polio's debilitations. You were now told not "use it or lose it" but, instead, "use it and lose it." For some of us, following this advice meant, or would mean, the end of our lives as we were living them.

"Be active for 10–15 minutes, then rest for 5"—yeah, right, I remember thinking: that would really work while one is rushing down an airport concourse deploying the raised handle of one's "rolly bag" as a crutch. Told to always use wheelchair assistance for air travel, my reaction, again, was, "Yeah, right! Have you traveled lately and discovered how erratic are airline efforts to get persons with disabilities to their connecting flights—especially given that, although you may be the first to board, you are often the last to deplane, since the airline doesn't want to slow other passengers down, as would be the case if disabled passengers got off first and into the wheelchair (assuming, of course, that the wheelchair was even there in the first place)?"

What follows is something of a digression, but it may help to illustrate the reality of life for traveling post-polio persons. Assume you have asked for wheelchair assist. You find a rather bedraggled and, no doubt, underpaid wheelchair assist person who pushes the passenger—or two passengers—plus luggage up the jetway. (One wonders, as this process goes forward, why it is that jet bridges invariably *ascend*?) All polio people collect stories of what it means in practice to be "helped". Here are two of my own personal favorites.

In a northern England airport, I find myself abandoned by one wheelchair assist person with the cheery assurance that another will come along "shortly," only to find myself waiting 10, then 15 minutes *outside* the airport building itself in a bitterly cold wind and faced with a locked door back into the airport. I knew if I tried to break back into the building all the security alerts would jump from orange to red, or whatever the British equivalent is, and I would be wrestled to the ground by agents. Finally, someone shows up and tears off at a sprint to get me to my regional airliner on time.

Very mysterious, to be sure, but that adventure lacks the flair exhibited at De Gaulle Airport in Paris. I had put in for wheelchair assistance, since De Gaulle is one of many airports from hell (as anyone who has flown through it realizes). We landed. No jet bridge. We deplaned by stairs, with buses waiting. I asked the flight attendant what to do, since I had asked for assistance. She shrugged her shoulders. Relying on the "kindness of strangers," like Tennessee Williams' Blanche Dubois, I enlisted a tall, strapping fellow to lug my suitcase down the stairs while I struggled to keep my balance and hang on to my overloaded smaller bag, laden with books and papers. We reached the bottom, and I followed the crowd and boarded a bus to the airport proper. Before the doors closed, a frantic woman rushed toward me and queried whether I was the person who had requested an assist. "Yes." She then grabbed my suitcase and began marching back toward the stairs we had just come down as we deplaned the Boeing 767.

I followed, of course, as I never separate from my bag. She started lugging the bag back up the steps, ordering me to follow her. I protested that going "up" was more difficult for me than descending. No response. Although I could scarcely believe this French farce was actually being played out, I staggered after her, nearly stumbling and falling in the process. The lady was a house-a-fire. We reached the top. I was panting and instructed to stand there and wait. Five minutes later, an infernal contraption that looked like a stainless steel or tin cabin of very small proportions, was hoisted aloft and connected to the airplane's front exit opposite the exit from which we had deplaned. I was told to enter this windowless contraption. It was cold and dirty, trash strewn on the floor. The driver then disentangled the cabin from the plane and drove, juddering all the way, to an airport concourse bridge where another worker was poised to open a side exit. I then deplaned directly from the nasty contraption to the concourse.

What then? No one was waiting for me with a wheelchair. I was told to wait. Thanks, but no thanks! Let me the hell outta here! I began walking in the direction of what looked like an exit. Presently I heard a frantic voice behind me calling "Madame, Madame." I turned as a determined fellow, his long hair flying, caught up with me and nearly pushed me into the wheelchair where I clearly belonged. I am pleased to report that after all this silliness, I did connect up with a gracious graduate student who accompanied me on a train to Lyon, where I had a wonderful time.

The reader will, I hope, appreciate why post-polios who remain ambulatory may well opt for "thanks, but no thanks" when it comes to wheelchair assist. My view is: maybe, someday, it will be absolutely necessary. For now, I'll rely as much as possible on my own two legs, however "gimpy" and "limpy" they may be.

Back to the main narrative. . ..You see, being a polio hero, in your own eyes and that of the culture more generally—given the premium Americans place on individualism and self-help—meant *not asking for help*. I recall well the days when I was hauling around four children under the age of seven, grocery shopping, unloading all the groceries myself, often with the youngest in my arms, doing all the housecleaning, including scrubbing floors by hand—the only way to get them really clean according to my grandmother, who was never wrong about these things. All four children came with me to the laundromat much of the time, as we had neither washer nor dryer in those lean days. Sometimes I hauled the wet, heavy laundry home to hang it on clotheslines because that saved a few dollars. I did all the cooking. And I was studying for my Ph.D. at the same time.

Odd—even perverse—as it may sound, relinquishing these tasks over time was not an unalloyed relief to me. It also signified a failure of sorts: one was no longer completely "normal." We heroes defined our lives as "can do." Those who remained victims were "won't do," as, in many cases, they did not engage their own agency in an effort to get better. Now we were being told that our agency must push us in the direction of "should not do even if you can." We confronted a double-whammy: first our bodies did not permit us to do many things we had once done with ease and, second, as if that were not enough, we were being ordered, more or less, to relinquish even more. This stank.

15.3 Neither Victims Nor Heroes

I see a number of possibilities in light of these realities, including a cultural surround largely ignorant of the long polio story as it is a disease that has disappeared among us in the developed West—indeed, in most of the world. Unacknowledged as "triumph-over" by persons who have forgotten the left-alive polio people, the narrative nonetheless defines us or, better yet, yields a particular repertoire of possibilities.

Unsurprisingly, the first choice is to acknowledge the advice and the therapeutic wisdom of post-polio diagnoses but to say, "thanks, but no thanks," thereby keeping the heroic self-understanding, however battered and weakened, alive. Here the polio person takes a calculated risk that the cessation or radical abating of the powers that one draws upon to carry on, to live one's life, will not happen precipitously and all-at-once—and, when that happens, well, then one can deal with it.

A second response is to re-victimize oneself, however much the post-polio activists may speak of polio heroes. One thinks, "here it goes again, singling me out. Why me?" I am a victim. If the person has regained functions and capabilities, these are put on the shelf as the person may go beyond the post-polio recommendations to cease and desist. One returns to helplessness. I doubt there are many who inhabit this identity full bore but I have seen the victim dynamic in action, so the option is a reasonable one to identify.

The third option is to struggle to break out of the victim-hero scenario. One is no longer a child. The world is filled with good and ill. On some level, one is a victim. But you refuse to identify as such—you resist the blandishments of a popular culture that feasts on victimization, one of the least savory aspects of identity politics. On some level, you are also a hero. But even heroes slow down, get tired, grow old. You recognize that you have not been singled out for yet another travail over which you must triumph; rather, you confront, as all human beings must, limits, mortality, the loss of power. If you are a hero, you decide, it is of the quotidian sort, a heroism you share with thousands of ordinary people who both flourish and fail in their life-long pilgrimage. You may not be a polio kid anymore, but you have reason to feel pride at how you dealt with what was dealt you. No secure triumph-over, as you once believed, but neither did you capitulate. And that is enough.

Notes

1. There were, of course, adults stricken by polio, but the vast majority of polios were children—and that meant the disease was a particular focus of fear, horror, and sorrow.
2. That is, the victim could become hero, but the hero made sense only against the backdrop of victimization.
3. We know, e.g., that rape victims who refuse to see themselves solely through the lens of victimization that bids, in our current cultural moment, to become one's identity *tout court*, fare better over the long run than those who embrace the victim identity. Too, at least according to Bruno Bettelheim (1960), prisoners at Dachau who accepted that they had done nothing to deserve this fate and that no one was coming to remedy their particular situation; those who understood: I am caught in a demonic experiment for something I cannot help but be—a Jew or some other despised group—knew that their "job" for the duration was to figure out how best to survive. If one could not do this, one rapidly became a "Mussulman," one of the doomed or "drowned," in Primo Levi's (1959) terms.
4. I well recall my spinal puncture. My Daddy, as he was called then, was present and I was determined not to be a sissy in front of him—though he was not a stern person who insisted on such a standard of rectitude. My mother was less accepting of human weakness, so I'm sure that I had, by then, internalized her stoicism, to which was added my determination that my father remain proud of me, even in this screwed-up situation—and that staved off tears. I was on the road to rejecting victimization from that moment onward.
5. I recall a picture I drew in a letter I sent either to my Aunt Martha or Aunt Mary, of a "polio virus." That image, imagined by a 10-year-old, is in my mind's eye as I write these words. It goes without saying that this was the most hideous, menacing-looking bug one could imagine.

6. I recall that my dad, who had learned to do physical therapy (as had my mother), once asked me, "Jeanie, don't you want to stop now?" He had no doubt noticed the tears streaming down my cheeks as we did the stretching exercises. I wasn't actually "crying," for one's body protests what is happening to it and tears are shed even if you are not engaged in what is usually called "crying."

References

Bettelheim, B. (1960). *The informed heart: Autonomy in a mass age*. Glencoe, IL: Free Press.
Levi, P. (1959) *Survival in Auschwitz: The Nazi assault on humanity*. New York: Collier.

Index

Note: The letter 'n' following locators denote note numbers.

D.C. Ralston, J. Ho (eds.), *Philosophical Reflections on Disability*, Philosophy and
Medicine 104, DOI 10.1007/978-90-481-2477-0_BM2,
© Springer Science+Business Media B.V. 2010

LaVergne, TN USA
21 October 2009
161567LV00003B/72/P